Studies in Applied Philosophy, Epistemology and Rational Ethics

Volume 71

Studies in Applied Philosophy, Epistemology and Rational Ethics (SAPERE) publishes new developments and advances in all the fields of philosophy, epistemology, and ethics, bringing them together with a cluster of scientific disciplines and technological outcomes: ranging from computer science to life sciences, from economics, law, and education to engineering, logic, and mathematics, from medicine to physics, human sciences, and politics. The series aims at covering all the challenging philosophical and ethical themes of contemporary society, making them appropriately applicable to contemporary theoretical and practical problems, impasses, controversies, and conflicts.

Book proposals can be sent to the Series Editor, Prof. Lorenzo Magnani at lmagnani@unipv.it. Please include: A short synopsis of the book or the introduction chapter A Table of Contents The CV of the lead author/editor(s) For more information, please contact the Editor-in-Chief at lmagnani@unipv.it

Our scientific and technological era has offered "new" topics to all areas of philosophy and ethics – for instance concerning scientific rationality, creativity, human and artificial intelligence, social and folk epistemology, ordinary reasoning, cognitive niches and cultural evolution, ecological crisis, ecologically situated rationality, consciousness, freedom and responsibility, human identity and uniqueness, cooperation, altruism, intersubjectivity and empathy, spirituality, violence. The impact of such topics has been mainly undermined by contemporary cultural settings, whereas they should increase the demand of interdisciplinary applied knowledge and fresh and original understanding. In turn, traditional philosophical and ethical themes have been profoundly affected and transformed as well: they should be further examined as embedded and applied within their scientific and technological environments so to update their received and often old-fashioned disciplinary treatment and appeal. Applying philosophy individuates therefore a new research commitment for the 21st century, focused on the main problems of recent methodological, logical, epistemological, and cognitive aspects of modeling activities employed both in intellectual and scientific discovery, and in technological innovation, including the computational tools intertwined with such practices, to understand them in a wide and integrated perspective. **Studies in Applied Philosophy, Epistemology and Rational Ethics** means to demonstrate the contemporary practical relevance of this novel philosophical approach and thus to provide a home for monographs, lecture notes, selected contributions from specialized conferences and workshops as well as selected Ph.D. theses. The series welcomes contributions from philosophers as well as from scientists, engineers, and intellectuals interested in showing how applying philosophy can increase knowledge about our current world. Indexed by SCOPUS, zbMATH, SCImago, DBLP. All books published in the series are submitted for consideration in Web of Science.

Krystyna Bielecka

What is Misrepresentation?

The Role of Epistemic Accuracy in Cognition

Translated from Polish by Krystyna Bielecka and
Marcin Miłkowski

Springer

Krystyna Bielecka
Institute of Philosophy
University of Białystok
Białystok, Poland

Translated by
Krystyna Bielecka
Institute of Philosophy
University of Białystok
Białystok, Poland

Marcin Miłkowski
Institute of Philosophy and Sociology
Polish Academy of Sciences
Warsaw, Poland

ISSN 2192-6255 ISSN 2192-6263 (electronic)
Studies in Applied Philosophy, Epistemology and Rational Ethics
ISBN 978-3-031-84374-7 ISBN 978-3-031-84375-4 (eBook)
https://doi.org/10.1007/978-3-031-84375-4

The original submitted manuscript has been translated into English. The translation was done using artificial intelligence. A subsequent revision was performed by the author(s) to further refine the work and to ensure that the translation is appropriate concerning content and scientific correctness. It may, however, read stylistically different from a conventional translation.

Translation from the Polish language edition: "Błądzę, więc myślę: co to jest błędna reprezentacja?" by Krystyna Bielecka and Marcin Miłkowski, © Warszawa: Wydawnictwa Uniwersytetu Warszawskiego 2019. Published by Warszawa: Wydawnictwa Uniwersytetu Warszawskiego. All Rights Reserved.

This Springer imprint is published by the registered company Springer Nature Switzerland AG
The registered company address is: Gewerbestrasse 11, 6330 Cham, Switzerland

Acknowledgments This book is a revised translation of my monograph *Błądzę, więc myślę* (Wydawnictwa Uniwersytetu Warszawskiego, Warszawa 2019). Besides obvious changes and omissions addressed in the translation, it was enriched by including new work that was published since 2018. I would like to thank the anonymous reviewer for this series for their insightful remarks on revising the book, as well as Lorenzo Magnani, the series editor, for his clear and helpful comments.

Here, I wish to repeat the acknowledgments that appeared in the original Polish version. I extend my gratitude to Katarzyna Paprzycka-Hausman and Justyna Grudzińska for their substantive and helpful comments. I am deeply thankful to Andrzej Klawiter and Mieszko Tałasiewicz for their thorough reviews of the entire manuscript before it was published in Polish. I thank Marcin Poręba for his careful reading of Chaps. 3 and 7.

I am grateful to Robert Poczobut for the invitation to present Chap. 10 at the Cognitive Science Thursdays seminar series. My appreciation goes to Marcin Miłkowski for his substantive and editorial consultations, as well as for the invitation to present at the Cognitive Science Philosophy seminar, not to mention his encouragement to have my monograph published in the revised English translation. I appreciate the valuable consultations with Dr. Rafał Czajkowski on cognitive maps.

I extend my gratitude to Ruth Millikan for her insights on teleosemantics and biosemiotics, as well as our private correspondence. My thanks also go to Anton Markoš for consultations regarding biological examples and to Mark Bickhard for discussions on interactivism.

Finally, I wish to thank the participants of the "Cognition-Mind-Action" seminar, the Cognitive Science Philosophy seminar, the Cognitive Science Thursdays, and the "Gatherings in Biosemiotics" conference in Castiglioncello for their valuable input and discussions.

My work that led to this book was funded by the National Science Centre, under the decision DEC-2012/05/N/HS1/02900.

Competing Interests The authors have no competing interests to declare that are relevant to the content of this manuscript.

Contents

List of Figures

Chapter 1
Introduction

The question of the nature and role of mental representations is central to the philosophy of mind, as well as psychology and the broader cognitive sciences. The concept of mental representation is ubiquitously employed and serves various functions. We invoke representations in everyday language when discussing someone's thoughts or decisions, and in scientific research when investigating cognitive processes and their impact on behavior.

1.1 Representation, Error, and Error for the System

In contemporary philosophical discourse on the nature of representations, the fundamental questions are: What are representations? What constitutes the content of a representation? And by what means do representations acquire reference and content? The latter question is intertwined with the nature of intentionality, or aboutness. Aboutness is an essential characteristic of representations. They can be about actual states of affairs or about fictional entities. For instance, I can contemplate both a doctoral thesis and a vacation. We say that representations have content because they describe an object in some manner; they also have reference when they relate to an object that exists.

Intentionality is connected to the accuracy of representations. Representations may describe their targets either adequately or inadequately. Indeed, my thought that the sun is shining today is accurate when sunlight streams through the window, whereas the belief that today is Friday is mistaken if it is actually Thursday. Since representations can be correct or incorrect, the concept of representation is normative.

The claim that representations are normative might be challenged by appealing to simplicity: a non-normative concept of representation would be simpler. Yet, this approach has a significant flaw—it compels us to accept that all representations must invariably be accurate, or they do not qualify as representations. This, however, leads

K. Bielecka, *What is Misrepresentation?*, Studies in Applied Philosophy, Epistemology and Rational Ethics 71, https://doi.org/10.1007/978-3-031-84375-4_1

to highly problematic consequences. The notion of representation is tied to instrumental rationality when applied to describing human and animal behavior in intentional terms. Instrumental rationality involves choosing the right means to achieve a predetermined goal [1].

Consider Johnny, who intends to take a bus to the university, and examine two scenarios. In the first scenario, Johnny acts according to the bus schedule to reach the university, thereby selecting the appropriate means to his intended end. In the second scenario, Johnny recalls an outdated bus schedule. Unaware that the bus no longer arrives at that time according to the current schedule, he continues to wait for the bus.

Without the concept of mental representation—and particularly that of misrepresentation, exemplified by Johnny's erroneous belief that the bus will arrive at 7:45—explaining a student's behavior as rational becomes exceedingly difficult. Johnny engages in an action that fails to achieve his intended goal: he waits, even though the bus will not arrive. Eradicating the possibility of possessing representations, including erroneous ones, would thus result in the undesirable consequence that any objectively incorrect action would be indicative of irrationality.[1] Given that errors in action are commonplace, this would lead to the widespread ascription of irrationality. However, we can maintain the concept of rationality in cases of error if we employ the notion of misrepresentation. Johnny, who holds an incorrect representation of the bus's arrival time, acts rationally because he has reasons to believe the bus will arrive at that time.

Similar reasoning can be applied to animals. In 1932, Edward Tolman conducted an experiment in which a rat learned the layout of a maze's arms without the presence of food that could serve as a reward. Yet, when food was placed in the maze, the rat promptly located it, despite not being hungry [3]. Moreover, the rat exhibited surprise when it followed the same route that had previously led to food, only to find no food there this time. In the first case, the rat accurately represents the path to reach the location where food was previously found. One might say that the rat selects the appropriate means to achieve its intended goal.[2] In the second case, the rat fails to reach the intended goal: it heads toward the place where food used to be, but is no longer present. Without the concept of mental representation, exemplified by the rat's erroneous belief that food is located in a specific place, attributing rationality to the animal would be quite challenging. The concept of representation is crucial for explaining why the rat heads toward the former food location and, most importantly, why it is surprised when the food is not there.[3]

---

[1] Donald Davidson in his *Rational Animals* [2] argues that the irrationality of someone's beliefs or actions can only be attributed on the condition that they possess predominantly rational beliefs.

[2] Tolman defended a kind of behaviorism, which he called "purposive behaviorism" [4].

[3] William Bechtel [5] notes that without the representational framework, it would have been impossible to predict new intriguing phenomena or to design novel experiments. Tolman's more robust hypothesis possesses greater heuristic value than non-representational alternatives, and has sparked a still-vibrant research program. In 2014, the Nobel Prize was awarded to John O'Keefe and the Mosers for their discoveries within this research program.

We routinely employ the notion of misrepresentation when we describe and explain our behaviors that lead to our intended goals. Johnny stands at the bus stop at a given time because he is waiting for the bus; however, he harbors a mistaken belief about the bus schedule and is unaware that the bus will not come. The concept of mental representation also allows us to predict and change someone's behavior. Thus, we anticipate Johnny's actions by knowing his habits and, upon asking him where he is headed at a certain time, we expect him to be at the bus stop, intending to travel to the university. We indirectly influence Johnny's behavior by either restraining it or causing it to change when we alter his mental representations, such as by informing him that he is consulting an outdated timetable. It is noteworthy that by affecting someone's mental representations, we can alter their behavior; it would take considerably more effort to forcibly bring Johnny to the bus stop, assuming, charitably, that he is not exceedingly obstinate.

In summary, without the concept of misrepresentation, one cannot systematically explain behavior in terms of instrumental rationality, as it would require the assumption that rational cognitive systems universally err in selecting inappropriate means to their ends for unknown reasons. Furthermore, such purported irrationality in behavior could not be rectified by refining the cognitive system's reasoning, because the system is presupposed to have only correct representations. The only feasible method of addressing this enigmatic irrationality would likely be neurological intervention. What else could be done if Johnny were fundamentally infallible? If misrepresentations are nonexistent, a cognitive system with mental representations must have them all correct (assuming it has any): if the system represents something, it does so without error. Given the implausibility of human infallibility, I dismiss the suggestion to expunge the concept of misrepresentation from the theory of mental representations. This is precisely the type of undesirable outcome that Mark Perlman's approach, which advocates for the elimination of this concept, leads to [6].

The concept of misrepresentation plays a role not only in describing, explaining, and predicting seemingly irrational behaviors, but also in (at least certain types of) learning from errors. Such learning presupposes that an organism can recognize its own mistake. Suppose Stan is working through arithmetic problems from a collection whose answers are at the end of the book. Aware of an error in his calculations, Stan strives to identify and correct the cause, attempting to align his solution with the correct answer. Learning from mistakes is a widespread phenomenon, evident even in more basic motor skills. For instance, a child incorrectly riding a bicycle follows an improper motor plan, resulting in a fall. Error-based learning involves mastering the correct sequence of movements, which eventually prevents the child from falling.

These examples illustrate a broader principle: mental representations, including misrepresentations, are essential for explaining instrumental rationality and cognitive functioning across various levels of complexity. Our analysis demonstrates that the need for representation spans the full spectrum of cognitive processes, from basic error detection to sophisticated belief revision [7]. This view unifies our understanding of how cognitive systems rely on representational mechanisms to adapt and improve performance. By grounding representation in the requirements of rationality and learning, we provide a cohesive framework that explains how representations underpin our ability to learn from errors and navigate the world effectively, bridging cognitive and epistemological concerns.

1.2 Philosophical Assumptions

In this work, I adopt naturalistic assumptions.

Naturalism posits that (1) there exists solely a natural reality, unaffected by any supernatural entities, and thus (2) all aspects of reality are amenable to scientific investigation. The enumerated tenets of naturalism correspond to its two facets, leading to the distinction of its subtypes: (1) ontological naturalism and (2) methodological naturalism [8]. In this book, I adopt the stance of methodological naturalism, a prevalent position in contemporary philosophy, particularly within the philosophy of science. Methodological naturalism recognizes only the methods of contemporary science as valid for substantiating claims, along with any reliable and intersubjectively verifiable methods applicable in scientific inquiry, the outcomes of which are justifiable through these methods.[4]

Methodological naturalism need not be confined to the methods of established and mature sciences; as long as the proposed research method is shown to be consistent with the set of practices already recognized as valid in scientific inquiry, then an explanation employing this method aligns with methodological naturalism. Given that philosophy often spurs the development of research disciplines and suggests new investigative methods, philosophers frequently originate new scientific methodologies. For instance, William James, in the realm of psychology, advocated for the study of emotions through the examination of bodily states, drawing on his philosophical arguments.[5]

Methodological antinaturalism posits that certain fields of inquiry, such as the humanities, are treated distinctly from science, possessing different research objectives and methods. An instance of non-naturalistic methods in the humanities is empathy in the sense of Wilhelm Dilthey [11].[6]

---

[4] Determining what reliable and intersubjectively verifiable methods are is a very complicated matter. Methodological naturalism assumes that the reliability and intersubjectivity of research methods are subjects of examination within science itself, which adopts new methods using previously accepted ones. Science constantly extends and changes its repertoire of methods, which makes it comparable to the reconstruction of a boat at sea, according to the well-known metaphor of Otto Neurath [9].

[5] His argument unfolds as a thought experiment involving the removal of bodily aspects from the experience of emotions. The outcome of this experiment revealed that once the bodily aspects of emotional experience are removed, nothing remains. "If we fancy some strong emotion, and then try to abstract from our consciousness of it all the feelings of its characteristic bodily symptoms, we find we have nothing left behind (...)" [10, p. 193]. This line of reasoning enables psychologists to measure emotions by assessing the body's state, exemplifying a research methodology.

[6] It is worth noting that contemporary naturalism differs from nineteenth century naturalism, in which the cultural aspects of meaning phenomena were overlooked and reduced to purely biological or economic phenomena. Methodological naturalism does not preclude considering reasons as

Consider the naturalistic premise in the context of representations. Franz Brentano maintained that intentionality is the characteristic that distinguishes minds from the physical world [14]. His work revived the concept of intentionality in contemporary philosophy. Brentano defined "intentionality" as the reference to a certain content, direction toward a particular object, or immanent objecthood. For this philosopher, intentionality is specific to what is exclusively mental, as opposed to what is non-mental. In other words, intentionality is set in opposition to what pertains to the material world.

This dualistic restriction of the concept of intentionality to the mental is incompatible with the widely accepted naturalistic assumption in cognitive science research that intentionality is not fundamental in the way that concepts of basic and universal physical theories are. From a naturalistic standpoint, explicating the concept of intentionality requires concepts other than intentionality itself, namely, those already employed in science or those that could be explicated within these concepts. Similarly, explicating the concept of mental representation in non-representational terms is both necessary and warranted.

My commitment to methodological naturalism leads to:

Meta-desideratum (I): A satisfactory theory of representation is connected with empirical research.

Meta-desideratum (I) stipulates that a theory of representation should align with the findings and methodologies of empirical sciences, ensuring that the theory's assumptions do not deviate from those held in empirical sciences.

Theories of representation often struggle to correspond with empirical research. Non-reductive naturalists like Searle [15], Mendelovici [16], Mendelovici and Bourget [17], Strawson [60], treat intentionality or phenomenal consciousness as conceptually primitive. This approach is contentious because it doesn't stem from findings or assumptions in empirical research, potentially disconnecting the philosophical theory from scientific practice.

Burge [18] is more in tune with assumptions in psychology. He claims representation is conceptually primitive, stating that "trying to reduce representation and

---

causes of behavior, notably in classical economics [12], and thus aligns with the anti-positivist naturalism of the Poznań Methodological School [13], pp. 175–182]. Anti-positivist naturalism acknowledges the existence of idealizing concepts ("typically-ideal") in both the humanities and natural sciences, leading to the repudiation of positivist inductionism in both scientific domains. The stance against inductionism and the assertion that humanistic interpretation has a rationalizing character mean that anti-positivist naturalism rejects not only inductionism but also phenomenalism and psychologism. As such, it stands in stark contrast to positivist naturalism.

In contrast to the anti-positivist naturalists of the Poznań School, methodological naturalists permit neuropsychological or behavioral research into motivations. Moreover, unlike the anti-naturalists of the Poznań School, methodological naturalists typically do not acknowledge the autonomy of the humanities. However, methodological naturalism does not necessarily lead to extreme biological or neuropsychological reductionism. Reductionism in the humanities contends that all cultural and semantic phenomena can be reduced to biology or neuropsychology, whereas extreme reductionism argues that some terms denoting cultural or semantic phenomena are replaceable with biological or neuropsychological terms, and others should be eliminated from theoretical discourse.

veridicality to something more 'naturalistically acceptable' is probably pointless and hopeless" (p. 298). However, this stance still requires substantial philosophical work to unpack and risks reaping the benefits of theft over honest toil by leaving representation as primitive.

In contrast, I propose a reductionist program that adheres closely to empirical research. This approach aims to bridge the gap between philosophical theories of representation and scientific investigation, avoiding the introduction of primitives without empirical grounding. By maintaining a strict connection to empirical findings, this program offers a more promising path towards a naturalistic understanding of representation, addressing the limitations found in both non-reductive and primitivist theories.

Focusing on mechanisms is part of the ongoing cognitive (neuro)science revolution, providing a framework for understanding how representation can be naturally accounted for [19]. This approach explains complex phenomena through the interactions and organization of their parts, offering an empirical basis for reducing representations to more fundamental processes [20].

A theory of representation should also clarify the nature of representational intentionality. Only then will the scientific notion of mental representation align with its practical usage. Absent this premise, even an extreme reductionist view of intentionality, such as framing it as a causal relation, would be wholly adequate. Hence, I propose the following meta-desideratum:

Meta-Desideratum (II): A satisfactory theory of representation can explicate the representational nature of intentional concepts in both everyday and scientific practice.

By "explication," I refer to a procedure aimed at conceptual analysis (distinct from explanations that concern certain facts—explained by other facts, such as natural laws or causes). Therefore, through the conceptual analysis of representational concepts, bolstered by examples from scientific practice, the essence of their representational quality should become evident.

Thus, the concept of phenomenal consciousness fails to adequately explicate the concept of intentionality, as the concept of phenomenal consciousness is in these theories equally or even more obscure than that of intentionality (but see Mendolovici and Bourget [17]).

The second philosophical premise of the book is the commitment to realism regarding the existence of mental representations. Realism is at odds with anti-realism: anti-representationalism, eliminativism, and instrumentalism. Ontological anti-representationalism [21] asserts that representations do not exist; epistemic anti-representationalism [22] claims they serve no explanatory function. A particular variant of anti-representationalism is eliminativism [23], which argues that some representations can and should be eliminated from a future, comprehensive psychological or neurocognitive theory. Instrumentalism (as ascribed to Dennett [24]), on the other hand, contends that the concept of representation serves descriptive, predictive, and explanatory purposes, yet it need not be employed with the presupposition that it refers to anything real.

Contrary to ontological anti-representationalism, realism posits the existence of representations; against eliminativism and epistemic anti-representationalism, it asserts that these representations serve descriptive and explanatory functions in mature cognitive theories; and against instrumentalism, it claims that the term actually refers to certain entities with specific causal roles in cognitive systems.

In this book, I argue that one can identify common features of the concept of mental representation as it appears in both everyday practice and scientific inquiry. At times, it is only evident in the form of detailed intentional concepts, such as "thought" or "perception.“ I delineate the characteristic features of a prototypical concept of representation. I refrain from speculating on the types of representations that exist, their nature, or their differences. I do not presuppose that the prototypical concept of mental representation is synonymous with the philosophical concepts of belief or desire; however, I posit that it shares at least some of their features. I also posit that if beliefs and desires do exist, they are forms of representations. In contrast to the classical view of mental representation, I do not confine myself to representations such as beliefs and desires that are linguistically expressible, as Donald Davidson, for example, does.

In the discussion of mental representations, the naturalization of intentionality and its associated naturalization of representational content pose particular challenges for representationalists [25]. Therefore, this work concentrates on the naturalization of intentionality. It is the proponents of representationalism, such as Ned Block, Fred Dretske, and Ruth Garrett Millikan, who undertake the task of explicating intentionality in non-intentional terms, and their approaches will be examined in the book.

I exclude two types of positions from my analysis. First, those that assume realism regarding intentionality but anti-realism concerning mental representations. Daniel Hutto and Erik Myin's position is one such example. In contrast, my work addresses a realistic interpretation of intentional states and mental representations. While I do not examine Hutto and Myin's anti-representationalism in detail, in particular their later work [26], I deal with the so-called "hard problem of content" they present, which forms a compelling argument against realism. Second, I set aside positions that accept both intentionality and realism in terms of mental representations but do not naturalize intentionality. Searle's theory [15] is an example of this.

I argue for a philosophical conception of mental representations that could serve as a meta-theory for empirical theories of mental representations. The philosophical conception alone is not sufficient to meet the proposed desiderata, as it also requires the construction of an empirically adequate theory or model. Since I adopt the assumptions of methodological naturalism, this conception should meet these desiderata by appropriately describing the structure of phenomena related to representations. For simplicity, I will also refer to philosophical conceptions of representation in the book as meeting or not meeting certain desiderata; this will mean that an empirical theory or model using such a conception would meet my desiderata. The philosophical conception alone cannot literally fulfill them; it can only provide the appropriate conceptual tools. I do not provide an empirical theory of mental representations, as that is not the role of a philosopher.

The aim of this book is to justify and organize the requirements that a satisfactory philosophical account of mental representations should meet, which are consistent with common practice and scientific research.[7] I analyze selected contemporary conceptions of mental representations in terms of the introduced meta-desiderata and corresponding desiderata, and discuss the requirements for philosophical accounts of various types of mental representations. As a result, I develop my own naturalistic account of mental representations that, in my opinion, meets all the desiderata and provides a proper account of representational error.

One of the main problems of the conception of representation is the account of representational error. Not all contemporary proposed conceptions address this problem, and even if they do, they do not develop an account of error for the system. This book provides an original conceptualization of representational error detectable by the cognitive system within the framework of teleosemantic theory.

The book is methodological and theoretical in nature. I rely on conceptual analysis. Conceptual analysis is usually understood as providing the necessary and sufficient conditions for the application of a given term, thereby constructing an explicit definition of the term, which might be a synonymous definition or an equivalence definition. In the book, I do not provide a normal definition of the concept of mental representation, but rather its implicit definition.[8] I compile a list of selected requirements for a theory of mental representations, which I call "desiderata for an adequate theory of mental representation." I justify and organize the selected desiderata.

Due to the role of examples, the conceptual analysis conducted in this book resembles a 3-analysis in the sense of Paul Thagard [62], who distinguishes three elements of such analysis: (1) providing prototypical examples, (2) indicating typical features, (3) describing characteristic explanations using the given concept. On the one hand, I examine examples from folk practice and scientific research, and on the other hand, I select only certain uses of the concept of mental representation as those that *prima facie* meet the desiderata I have identified for the concept of representation. I take paradigmatic examples of the use of the concept of mental representation seriously precisely because I adopt a naturalistic stance, according to which philosophy does not have a privileged point of view that allows it to resolve all theoretical disputes, but must refer to what is currently considered best justified. This is the characteristic of the paradigmatic examples I cite. Philosophy should not

---

[7] One can imagine at least two ways of achieving this goal. The first would involve creating a meta-theory, i.e., formulating a priori requirements for a philosophical account of mental representations, and then organizing and justifying them. This is not my intention. Another way to achieve this goal is post hoc: it involves formulating desiderata, organizing and justifying them as a result of a systematic review of contemporary philosophical accounts of mental representations, assuming naturalism and realism. This review provides both an argument for such an account that would meet the specified desiderata. In this sense, creating such an account involves justifying and organizing the requirements that a satisfactory philosophical theory of mental representations should meet.

[8] As argued, for example, by Fodor [27], we are unable to provide explicit definitions, i.e., intensional definitions, of elementary empirical concepts. Of course, there are intensional definitions of mathematical concepts (e.g., "A square is a rectangle with all sides equal").

only formulate the norms governing the use of the concept of representation but also correctly describe its use.

I analyze examples to further justify why I narrow the concept of mental representation in this particular way. By analyzing examples, I point out the specific role that mental representations have—just as the concept of moral good has its indispensable application in ethics to describe and explain a class of positive values, so the concept of mental representation has its indispensable application in philosophy and cognitive science, as it allows distinguishing a certain class of cognitive processes. In other words, I show in what sense the concept of representation is not trivial, which allows meeting Ramsey's job description challenge [28]. Using selected examples from the sciences of the mind and brain, I demonstrate that it is possible to speak of a certain unified concept of mental representation in humans and animals that meets the desiderata I have identified.

The desiderata and meta-desiderata I propose should be met by empirically testable theories or models[9] utilizing the concept of mental representation. Since all such models and theories rely on a certain theoretical apparatus, meeting the desiderata will also depend on whether this apparatus is methodologically consistent with them. For example, Skinner's behaviorism [29] does not meet any of my desiderata: behavior is explained only in terms of stimuli and responses, without postulating any structure of the cognitive system, and thus without representations.

Meta-desiderata lead to increasingly detailed desiderata. I will now provide a justification for the proposed list. A more comprehensive justification is provided when discussing individual philosophical conceptions and analyzing examples from mind and brain sciences.

I argue that the concept of mental representation should play various roles, such as describing, explaining, and predicting behaviors. I justify this by stating that only such a concept of mental representation is consistent with common practice and scientific research. Additionally, I note that by changing intentional states, one can change the behaviors of cognitive systems, i.e., intervene in their behavior. In this regard, I formulate the explanatory desideratum:

> I.1. A satisfactory theory of representation takes into account the role of representational concepts in describing, explaining, predicting, and controlling cognitive processes and behaviors.

The theory of representation should encompass the broadest range of cognitive processes and behaviors, in such a way that it takes into account the specificity of representational phenomena and at the same time does not lead to the trivialization of the concept of mental representation. The concept of representation should not be a priori restricted only to human behaviors or only to animal behaviors, nor only to behaviors considered at the personal level or only to processes occurring at the

---

[9] I use the terms "theory" and "model" as they are employed in cognitive science. A theory describes a class of phenomena (e.g., those explained or predicted). A model, on the other hand, describes specific phenomena or a group of specific phenomena, typically utilizing a more general theory.

subpersonal level. This idea is expressed by the universality desideratum (I.2) and the multi-level desideratum (I.3):

I.2. A satisfactory theory of representation concerns both linguistic and non-linguistic (e.g., perceptual) phenomena in humans and animals.
I.3. A satisfactory theory of representation concerns phenomena considered at both the personal and subpersonal levels.

These desiderata are justified by the fact that humans are also animals, therefore an adequate theory of representation should take into account not only human behaviors and processes, and particularly not only cognitive processes related to language.

The concept of mental representation is associated with rationality. In the biological world, manifestations of instrumental rationality are encountered. I frame this as the rationality desideratum:

II.1. A satisfactory theory of representation is consistent with assumptions regarding the rationality of cognitive systems.

In the book, I do not provide a specific theory of rationality nor do I explicate the relationship between rationality and representation in detail. However, I argue that a theory of mental representations that describes the behaviors of humans and animals should not imply that a cognitive system making errors is entirely irrational.

The concept of mental representation is related to the reflection of reality. This has a biological justification, as living solely in a world of fiction or imagination would not allow organisms to survive. As such, the concept of mental representation should not pertain exclusively to empty or fictional representations. This idea is expressed by the desideratum of the naturalization of intentionality, which breaks down into the desiderata of the determination of reference and content and the desideratum of content bearing:

II.2. A satisfactory theory of representation naturalistically explicates the concept of intentionality, which means:
II.2.1. It identifies the determinants of reference and content.
II.2.2. It explains why a representation can bear content.

In classical philosophical conceptions, such as those of John Stuart Mill or Gottlob Frege, representations have both reference and content. A concept of mental representation that considers only reference can designate only simple objects or properties. The fact that representations can bear content is a useful feature for organisms. Content consists of satisfaction conditions, which in the case of more complex declarative linguistic representations are identical to truth conditions. If representations could not have satisfaction conditions, they could not be evaluated, and thus organisms could not infer the cognitive value of the represented objects or properties. A conception of mental representations that does not meet the desideratum of content bearing, and thus does not naturalize the concept of content, does not solve the so-called hard problem of content [25]. According to Hutto and Myin, the problem is hard because, in naturalistic conceptions of intentionality, explicating the concept of satisfaction conditions is problematic.

Productivity is a property of the structure of the representation system, enabling the generation of potentially infinitely many representations. According to Fodor, the only credible explanation for the productivity of the representation system is its compositionality and systematicity. If a representation system is compositional, representations combine into more complex representations or decompose into simpler ones. Unlike compositionality, the concept of systematicity is introduced only by recourse to intuition [30]. Thus, a representation system is systematic if it can generate both the representation "John loves Mary" and "Mary loves John." Systematicity is therefore a property of the representation system that allows for the creation of representations in such a way that if a representation with a certain structure can be formed, a representation with another, systematically altered structure could also be formed. Unfortunately, the catalog of these changes, which would allow for the specification of this somewhat vague definition, has not been precisely determined in the literature. In summary, the condition for a representation system to be productive is systematicity and compositionality.

A satisfactory theory of representation should provide an appropriate conception of content, expressed by the desideratum of an appropriate conception of content (II.3) and the subordinate desiderata of productivity (II.3.1), as well as two desiderata concerning the consideration of error (namely, II.3.2 on the possibility of error and II.3.3 on the system-detectable error):

> II.3. A satisfactory theory of representation provides an adequate concept of content (including the non-contingent relationship between intension and extension).
> II.3.1. A satisfactory theory of content of representation takes into account the productivity of component parts of representation.
> II.3.2. A satisfactory theory of the content of representation accounts for the possibility of misrepresentation.
> II.3.3. A satisfactory theory of the content of representation accounts for the capacity of the cognitive system to recognize the accuracy of the representation.

I justify the desideratum of productivity by stating that a satisfactory theory of mental representation should encompass the structural properties of representations, which make some of them productive. This particularly applies to conceptual and linguistic representations [30]. I justify the desideratum of the possibility of error by stating that considering the possibility of misrepresentation allows one to distinguish misrepresentation from the absence of representation. A satisfactory mental representation should be valuable to the cognitive system. This occurs when it has a cognitive character for that system, meaning the organism is able to distinguish (at least some) misrepresentations from correct ones. Therefore, I adopt the desideratum of the system-detectable error—according to this desideratum, a satisfactory theory of content should consider the possibility of the cognitive system recognizing the accuracy of the representation. Both requirements are crucial because they indicate that cognitive systems can make errors through representations, the desideratum of system-detectability of error is particularly important because it speaks to the specifically representational role of mental representations.

In the first part, I start from the observation that intentional concepts, including beliefs and desires, but not limited to them, are widely used in common practice. I present Dennett's conception of the intentional stance, as he defends the broad use of such concepts. I also draw attention to the compatibility of explanations in terms of mental representations with the principles of rationality and charity. I thus regard the conception of the intentional stance as a correct philosophical analysis of common practice.

In this chapter, I also demonstrate how intentional concepts are related to rationality. Remaining consistent with Dennett's conception of stances, I challenge Davidson's argument against animal rationality. The outcome of Chap. 2 is the defense of the concept of mental representation as an intentional concept, consistent with common practice and scientific research. In this chapter, I primarily argue for the desiderata: explanatory (I.1), universality (I.2), and multi-level (I.3), as well as rationality (II.1).

In the second part, I focus on the simplest conceptions of mental representations, which concern not only the level of entire organisms but also subpersonal cognitive processes. I examine selected functionalist positions. In this chapter, I discuss a purely internalist conception. I show that internalism has certain advantages, as it emphasizes the internal dimension of the cognitive system, focusing on the intensional characterization of representations. I accept some explanatory reasons for the narrow content conception (following Fodor and Segal), showing why the concept of content is important in theories of mental representation, but I argue that the concept of narrow content is not suitable for explicating concepts used in mind and brain sciences. Therefore, in the rest of the book, I limit myself to externalist conceptions of mental representation.

In Chaps. 5 and 6, I analyze causal-informational conceptions that focus exclusively on reference while ignoring content. These include: the Wisconsin-style causal semantics and Davidson's truth-conditional semantics. I show that these conceptions are limited to reference or truth conditions, which involve conceptual difficulties.

In summary of the second part, I state that none of the simplified theories of mental representation—neither extreme internalism nor extreme externalism—accounts for the common usage of the concept of mental representations, which demands considering both the reality outside the mind and the internal perspective of the cognitive system. They also do not account for the relationship of mental representations with scientific research. Thus, the lack of content, reference, or veridicality does not allow for the use of common explanations in mind and brain sciences using cognitive maps and concepts. Therefore, in line with externalist theories, I state that theories of mental representations should consider reference and veridicality to meet the desideratum of naturalizing intentionality.

I also show that a purely externalist theory of mental representation has conceptual difficulties with determining reference, which it cannot overcome without the concept of content. I argue for the necessity of considering both the content factor and the reference factor (thereby enabling the theory to meet the desideratum of the possibility of error). Block's two-factor conception serves as an example of an improper relationship between extension and intension, demonstrating that this relationship cannot consist solely of the conjunction of separate theories of content and reference.

Only externalist teleosemantic conceptions (from the third part) appropriately grasp the relationship between content and reference, as they provide the concept of wide content. Teleosemantics justifies the relationship between content and reference on the basis of the conception of the biological function of cognitive systems. Biological systems use mental representations for their better orientation in the world, necessary for survival and adaptation to the environment. Most importantly, the teleosemantic conception of function allows for tying content to reference in such a way that it enables attributing content to representations that are either erroneous or correct, which is the most important element of a satisfactory theory of mental representations. Thus, teleosemantics fulfills the first of the key desiderata of the possibility of error.

It is worth emphasizing that I allow for the theoretical possibility of explicating representational error in non-functional terms. Currently, however, only functional teleosemantic theories provide a satisfactory explication of this error (Fodor's asymmetrical causal theory [31, 61], discussed in the second part, remains too broad, and its criterion of correctness is not specific to the theory of mental representation). This is because, in teleosemantics, concepts such as "representing," "misrepresenting," and "non-representing" are explicated respectively in terms of function, dysfunction, and lack of function.

At the same time, teleosemantic conceptions are unable to handle the explication of the idea of the system-detectable error (error for the system), thus they do not meet the second key desideratum of system-detectability of error. An exception is Mark Bickhard's theory, which, however, is limited to anticipatory mental representations and contains strong metaphysical assumptions of processualism. In the last chapter of the book, I propose an original coherence-based conception of error for the system, which does not require anticipation. The conception I propose is based on Millikan's model of representation and explicates the idea of system-detectable error in a new way.

1.3 Examples of the Application of the Concept of Mental Representation in the Sciences of Mind and Brain

In scientific practice, the concept of mental representation is employed not merely instrumentally but particularly when theorists strive to ascertain the factors that endow representations with their specific content. I will refer to three classic examples that describe behaviors and are recognized by scientists as uncontroversial instances of mental representations treated realistically (for a recent review of how mental representations are studied in neuroscience, see [32]. While in this section, I only briefly introduce them, sketching the historical background, particular experimental tasks and scenarios are discussed at greater length in later chapters.

The first example originates from the study of rodents and involves a structural representation (cognitive map), while the remaining two, from cognitive psychology (classical theory of concepts, debate over mental imagery), include one structural representation (mental imagery) and one symbolic representation (rombix). These selected examples not only provide additional support for the desiderata but also allow for the evaluation of the utility of content theories. In analyzing the use of concepts of mental representations through selected examples, I will refer to various experiments employing these concepts; the artificial concept of rombix (in original Polish, *klipiec*) is an exception, as it is inextricably linked with Andrzej Lewicki's pioneering research on concept learning.

Structural representations are characterized by a homomorphism between the structure of the representational vehicle and that of the referent—the syntax of the vehicle informs us about the features of the referent [28, 33–36]. For instance, a map of Warsaw that indicates a bus stop informs (or at least should inform) us about the location of that stop in relation to other objects depicted on the map.

Symbolic representations are marked by a certain arbitrariness in the relationship between the referent of the representation and its vehicle, such that the structure of the vehicle does not inform us about the referent. For example, a symbolic representation of a cow could be the word "krowa" in Polish or "cow" in English, or it could even be represented using the decimal system. However, regardless of the format of the representation, the structure of its vehicle does not inform us about the structure of the referent—in this case, the cow (a cow is not composed of letters or elements that would correspond to letters in any natural language).

1.3.1 Cognitive Maps

Today, the status of cognitive maps as structural representations no longer sparks the controversies it did until recently.

John O'Keefe discovered place cells in mammalian brains in the 1960s, while Edvard and May-Britt Moser identified grid cells three decades later. The former are located in the hippocampus, and the latter in the entorhinal cortex. Together, these two cell types create an orientation system akin to a cognitive map [37]. For their collaborative discovery, O'Keefe and the Mosers were awarded the Nobel Prize in 2014. The place and grid cells identified by the laureates constitute a cognitive map in the most literal sense—a dynamic structure that evolves over time, mirroring the geometry of the environment. More precisely, it is a temporal pattern of neuronal excitations geometrically linked, such that neuronal activity reflects the spatial structure of the surroundings.[10] Edward Tolman had earlier hypothesized the existence

[10] The term "cognitive map" is used in various senses [38]: trivial (the cognitive map is any mental or neuronal mechanism enabling an animal to orient in its environment), loose (the cognitive map

of a cognitive map [40]. Furthermore, the cognitive map hypothesis still plays a heuristic role in neuropsychological research [5].

The status of cognitive maps was, until recently, a contentious issue between proponents of behaviorism and representationalism. The concept of cognitive maps emerged from a critique of crude behaviorism, which relied on a stimulus–response paradigm. Tolman demonstrated that an animal (specifically, a rat) could learn without reinforcement.[11] John O'Keefe and Lynn Nadel's research has revealed that the cognitive map is neurologically based in the rat brain. It can be observed and pinpointed within the hippocampus [41]. Further studies have demonstrated the presence of both retrospective and prospective memory not only in rodents but also in crows [42]. Some of these studies focused on rat behavior in a water maze where the rat was tasked with locating food in complete darkness, relying solely on stimuli related to its own movement: the sense of balance and the motor command copies sent to its muscles [43]. When left in darkness, rats are adept at finding food, which substantiates the existence of episodic memory in animals.[12] In other words, processing environmental data is significantly more complex and cannot be convincingly described solely in terms of stimulus without reference to intricate memory processes, particularly when a simpler alternative explanation in terms of cognitive maps exists. Furthermore, the debate over cognitive maps sometimes presents a false dichotomy: behaviorists assert the importance of the stimulus, believing that proponents of mental maps trivialize its significance [46]. In reality, visual stimuli play a crucial role in the formation and shaping of cognitive maps, a fact not contested by advocates of mental maps.

In summary, the arguments supporting the existence of cognitive maps and their role in describing, explaining, and predicting behavior suggest that the cognitive map hypothesis is a strong contender against behaviorist explanations, which reduce animal orientation in their environment to stimulus–response learning [35].

Recent research on cognitive maps in rats posits that sensory stimuli are utilized for correcting errors in current or future locations [47]. Additionally, there is evidence

---

is a mental representation that reflects the geometric properties of the environment); and literal (the cognitive map is a mental representation possessing the same foundational properties and representational mechanisms as a conventional map). Psychologists do not dispute the trivial sense of the cognitive map, as it is tautologous that all animals capable of spatial orientation possess it. The subjects of genuine debate are the other two senses—the loose sense, wherein the cognitive map pertains to geometric properties, and the literal sense, wherein the cognitive map embodies geometric properties. The loose sense is agnostic regarding the vehicle of representation, allowing the map to be linguistic, pictorial, or structural in nature; the literal sense, on the other hand, mandates that the cognitive map be construed as a structural representation [38]. Empirical evidence supports the existence of cognitive maps in the loose sense among mammals [39].

[11] Tolman performed an experiment in which rats freely explored a maze, learning the layout of its arms, even though they did not find any reward or penalty there.

[12] In addition, there is empirical evidence of memory trace formation in a rat's brain, in which place cells located in the hippocampus play a main role. These studies demonstrate a fundamental link between cognitive maps and episodic memory [44, 45].

of an error-related negativity signal (ERN) in EEG readings [48, 49].[13] Another indication that mammals recognize their mistakes is the concurrent release of dopamine in their brains. This implies that rats are capable of not only misrepresenting their spatial position, but also recognizing and correcting this misrepresentation [50].

1.3.2 *Mental Imagery*

Mental imagery is considered a classic example of structural representation. The term "mental imagery" currently denotes a technical category of mental processes involving the appropriate transformation of the content of ongoing perceptual acts. An example might be mentally rotating a figure that is being visually observed. Mental imagery is at the heart of a debate that hinges on whether certain cognitive processes rely on structural representations akin to images or on symbolic representations and tacit knowledge, which includes understanding of language grammar, intuitive physics, and fundamental social norms [51]. It is also significant in accounting how we think in images [52].

In the famous experiment by Roger Shepard and Jacqueline Metzler [53], participants were shown projections of two three-dimensional figures on a screen and asked to determine whether the figures were identical but rotated relative to each other, or different with one being the mirror image of the other. Shepard and Metzler discovered that the response time required to reach a decision was linearly proportional to the angle of rotation from the original position.

The controversy over mental imagery pertains to the ontological status of mental images. Proponents of the pictorial stance, including Shepard and Metzler [53], as well as Kosslyn [54] and Pinker [55], argue that subjects perform the same mental transformations as they would on paper or with actual objects in space. Thus, they posit the existence of mental images as structural representations in cognitive abilities related to mentally rotating geometric figures and assessing distances in the environment. Opponents of the pictorial stance, such as Zenon Pylyshyn, contest the claim that the proportionality of response time to the distance and rotation angle of a figure, as asserted by pictorialists, confirms the presence of a structural representation [53].

Pylyshyn [56] offered an alternative explanation that posits symbolic representations with a linguistic character, linked to tacit knowledge: he argued that pictorial representations emerge from the processing of symbolic representations and are secondary to them.

The debate over mental imagery also intersects with the broader discourse on the autonomy of psychology. Pylyshyn advocated for this autonomy, asserting that

[13] It is an evoked potential component, i.e. a component of electrical potential recorded from a head surface after an appropriate stimulus has been triggered.

the theory of mental imagery cannot be validated or invalidated by any neurological studies. Kosslyn, conversely, invoked cognitive processes that include brain functions.

The controversy over mental imagery was cast as a false dichotomy: structural representations were set against symbolic representations, which presuppose tacit knowledge. It appears that both positions in the debate negate each other, whereas proponents of structural representations can consistently utilize tacit knowledge, reasoning, and computational processes in their explanations of image formation (as noted by Kosslyn [57]).

1.3.3 Rombix

The third example I examine pertains to symbolic representations and stems from the classical theory of concept learning. Andrzej Lewicki conducted research on the acquisition of artificial concepts, including the concept of an object called a "rombix" ("klipiec" in Polish, an artificial word).[14] The study was conducted as follows:

> In Lewicki's experiments, the experimental material consisted of multicolored cards of various shapes, each featuring a small figure in the center whose color and shape varied. The researcher presented a standard card, declaring it a "rombix. " Subsequently, another card was shown, and the subject was asked whether it represented a "rombix. " If the response was incorrect, the researcher provided the correct answer before moving on to the next card. After displaying the initial series of cards, the researcher revisited each one, asking the subject to determine whether it exemplified the concept. This first set was shown repeatedly until the subject could correctly identify all exemplars. The researcher then introduced a set of control cards—cards the subject had not previously encountered. This set included both exemplars of the concept and non-exemplars. Correct identifications indicated that subjects could generalize the concept to new instances. [58: 303][15]

The study's objective was to engender a concept within the subject, thereby forming a representation of rombix.

In the most succinct terms: a representation emerges when the subject infers which features are applicable to a given concept (positive abstraction) and which are not (negative abstraction). The ability to differentiate rombices from non-rombices serves as a test to verify understanding of the rombix representation.

While there may be doubts about whether categorization necessitates concepts, Lewicki posits that it indeed does. A representation manifests in the subject when they can discern certain items. In the example provided, these are figures characterized by a specific color and shape.

---

[14] Andrzej Lewicki (1910–1972) was a distinguished Polish psychologist renowned for establishing the first Department of Clinical Psychology in Poland, located in Poznań. His innovative work included the development of an experimental method for generating artificial concepts and the clinical experiment, which have had significant implications in psychology.

[15] All translations are mine, unless otherwise noted.

Specifically, the term in the experiment denoted a class of colored images, all similar in that they were squares with a centrally placed red, diagonal square (a red diamond), yet varied by background color [59: 17].

The notion of representation is an element in explaining a subject's ability to recognize rombices based on the concept's essential features, which forms the foundation for communication with individuals who utilize this concept. Thus, representation is instrumental in elucidating both behavior and communication.

References

1. Weber, M.: Economy and Society: A New Translation (Trans. Tribe, K.). Harvard University Press (2019). https://doi.org/10.4159/9780674240827
2. Davidson, D.: Rational animals. Dialectica **36**(4), 317–327 (1982). https://doi.org/10.1111/j.1746-8361.1982.tb01546.x
3. Tolman, E.C., Honzik, C.H.: Introduction and removal of reward, and maze performance in rats. Univ. Calif. Publ. Psychol. **4**, 257–275 (1930)
4. Tolman, E.C.: Purposive Behavior IN Animals and Men. University of California Press, Berkeley (1949)
5. Bechtel, W.: Investigating neural representations: The tale of place cells. Synthese **193**(5), 1287–1321 (2016). https://doi.org/10.1007/s11229-014-0480-8
6. Perlman, M.: Conceptual flux: mental representation, misrepresentation, and concept change. Kluwer Academic Publishers, Dordrecht–Boston (2000)
7. Bielecka, K., Miłkowski, M.: Representationalism and rationality: why mental representation is real. Synthese **203**(5), 161 (2024). https://doi.org/10.1007/s11229-024-04540-z
8. Papineau, D.: Naturalism. In: Zalta, E.N. (ed.) The Stanford Encyclopedia of Philosophy. Retrieved from http://plato.stanford.edu/archives/fall2015/entries/naturalism/ (2015)
9. Neurath, O.: Protocol statements. In: Cohen, R.S., Neurath, M. (eds.) Philosophical Papers 1913–1946, pp. 91–99. D. Reidel Publishing Company, Dordrecht (1983)
10. James, W.: What is an emotion? Mind **9**(34), 188–205 (1884). 10/ch377c
11. Dilthey, W.: Introduction to the human sciences, selected works. In: Makkreel, R.A., Rodi, F. (eds.) Edited and translated, with an Introduction, vol. 1. Princeton University Press, Princeton (1883/1989)
12. Ross, D.: Philosophy of Economics. Palgrave Macmillan, Basingstoke (2014)
13. Kmita, J.: Z metodologicznych problemów interpretacji humanistycznej. Państwowe Wydawnictwo Naukowe, Warszawa (1971)
14. Brentano, F.: Psychology from an Empirical Standpoint. Routledge, London (1874/1995)
15. Searle, J.R.: Intentionality: an Essay in the Philosophy of Mind. Cambridge University Press, Cambridge, New York (1983)
16. Mendelovici, A.: The Phenomenal Basis of Intentionality. Oxford University Press (2018)
17. Mendelovici, A., Bourget, D.: Facing up to the problem of intentionality. Philos. Perspect.Perspect. **37**(1), 228–247 (2023). https://doi.org/10.1111/phpe.12188
18. Burge, T.: Origins of Objectivity. Oxford (2010)
19. Bielecka, K., Miłkowski, M.: Error detection and representational mechanisms. In: Smortchkova, J., Dołęga, K., Schlicht, T. (eds.) What are Mental Representations?, pp. 287–313. Oxford University Press (2020)
20. Boone, W., Piccinini, G.: The cognitive neuroscience revolution. Synthese **193**(3), 1509–1534 (2016). https://doi.org/10.1007/s11229-015-0783-4
21. Hutto, D.D., Myin, E.: Exorcising action oriented representations: ridding cognitive science of its Nazgul. Adapt. Behav.Behav. **21**(3), 142–150 (2013). https://doi.org/10.1177/1059712313482684

22. Chemero, A.: Anti-representationalism and the dynamical stance. Philos. Sci. **67**(4), 625–647 (2000)
23. Churchland, P.M.: The Engine of Reason, the Seat of the Soul. MIT Press, Cambridge, Mass (1996)
24. Dennett, D.C.: Intentional systems. J. Philos. **68**(4), 87–106 (1971). https://doi.org/10.2307/2025382
25. Hutto, D.D., Myin, E.: Radicalizing Enactivism: Basic Minds Without Content. MIT Press, Cambridge (2013)
26. Hutto, D.D., & Myin, E.: Evolving Enactivism: Basic Minds Meet Content. MIT Press (2017)
27. Fodor, J.A.: Methodological solipsism considered as a research strategy in cognitive psychology. Behav. Brain Sci.. Brain Sci. **3**(1), 63–109 (1980). https://doi.org/10.1017/S0140525X00001771
28. Ramsey, W.M.: Representation Reconsidered. Cambridge University Press, Cambridge (2007). https://doi.org/10.1017/CBO9780511597954
29. Skinner, B.F.: About Behaviorism. Knopf, New York (1974)
30. Aizawa, K.: The Systematicity Arguments. Kluwer, Boston (2003)
31. Fodor, J.A.: A Theory of Content and Other Essays. MIT Press, Cambridge (1992)
32. Thomson, E., Piccinini, G.: Neural representations observed. Mind. Mach. **28**(1), 191–235 (2018). https://doi.org/10.1007/s11023-018-9459-4
33. Cummins, R., Roth, M.: Meaning and content in cognitive science. In: Schantz, R. (ed.), Prospects for Meaning, pp. 365–382. de Gruyter, Berlin, New York (2012)
34. Gładziejewski, P., Miłkowski, M.: Structural representations: causally relevant and different from detectors. Biol. Philos. **32**(3), 337–355 (2017). https://doi.org/10.1007/s10539-017-9562-6
35. Shea, N.: Representation in Cognitive Science. Oxford University Press (2018)
36. Sloman, A.: The Computer Revolution in Philosophy: Philosophy, Science, and Models of Mind. Humanities Press, Atlantic Highlands (1978)
37. Moser, E.I., Kropff, E., Moser, M.B.: Place cells, grid cells, and the brain's spatial representation system. Annu. Rev. Neurosci.. Rev. Neurosci. **31**, 69–89 (2008). https://doi.org/10.1146/annurev.neuro.31.061307.090723
38. Rescorla, M.: Cognitive maps and the language of thought. Br. J. Philos. Sci. **60**(2), 377–407 (2009). https://doi.org/10.1093/bjps/axp012
39. Cummins, R.: Representations, Targets, and Attitudes. MIT Press, Cambridge (1996)
40. Tolman, E.C.: Cognitive maps in rats and men. Psychol. Rev. **55**(4), 189–208 (1948)
41. O'Keefe, J., Nadel, L.: The Hippocampus as a Cognitive Map. Clarendon Press, Oxford, New York (1978)
42. Trojan, M.: Na tropie zwierzęcego umysłu. Wydawnictwo Naukowe, Warszawa (2013)
43. Cheung, A., Ball, D., Milford, M., Wyeth, G., Wiles, J.: Maintaining a cognitive map in darkness: the need to fuse boundary knowledge with path integration. In: Sporns, O. (ed.) PLoS Comput. Biol. **8**(8), e1002651 (2012). https://doi.org/10.1371/journal.pcbi.1002651
44. Eichenbaum, H.: Hippocampus: cognitive processes and neural representations that underlie declarative memory. Neuron **44**(1), 109–120 (2004). https://doi.org/10.1016/j.neuron.2004.08.028.
45. Eichenbaum, H., Cohen, N.J.: Can we reconcile the declarative memory and spatial navigation views on hippocampal function? Neuron **83**(4), 764–770 (2014). https://doi.org/10.1016/j.neuron.2014.07.032
46. Benhamou, S.: No evidence for cognitive mapping in rats. Anim. Behav.Behav. **52**, 201–212 (1996)
47. Ferbinteanu, J., Shapiro, M.L.: Prospective and retrospective memory coding in the hippocampus. Neuron **40**(6), 1227–1239 (2003). https://doi.org/10.1016/S0896-6273(03)00752-9
48. Fiorillo, C.D., Tobler, P.N., Schultz, W.: Discrete coding of reward probability and uncertainty by dopamine neurons. Science (New York, N.Y.) **299**(5614), 1898–1902 (2003). https://doi.org/10.1126/science.1077349

49. Valerio, S., Taube, J.S.: Path integration: how the head direction signal maintains and corrects spatial orientation. Nat. Neurosci.Neurosci. **15**(10), 1445–1453 (2012). https://doi.org/10.1038/nn.3215
50. Miłkowski, M.: Satisfaction conditions in anticipatory mechanisms. Biol. Philos. **30**(5), 709–728 (2015). https://doi.org/10.1007/s10539-015-9481-3
51. Thomas, N.: Are theories of imagery theories of imagination? An active perception approach to conscious mental content. Cogn. Sci.. Sci. **23**(2), 207–245 (1999)
52. Kozak, P.: Thinking in images: imagistic cognition and non-propositional content. Bloomsbury Academic, London, New York (2023)
53. Shepard, R., Metzler, J.: Mental rotation of three-dimensional objects. Science **171**(972), 701–703 (1971). https://doi.org/10.1126/science.171.3972.701
54. Kosslyn, S.M.: The Medium and the Message in Mental Imagery: a Theory. American Psychological Association, Washington (1981)
55. Pinker, S.: Mental Imagery and the Visual World. Harvard University Press, Cambridge (1980)
56. Pylyshyn, Z.W.: The imagery debate: analogue media versus tacit knowledge. Psychol. Rev. **88**(1), 16–45 (1981)
57. Kosslyn, S.M.: Image and Brain: the Resolution of the Imagery Debate. MIT Press, Cambridge (1994)
58. Maruszewski, T.: Psychologia poznania: sposoby rozumienia siebie i świata. Gdańskie Wydawnictwo Psychologiczne, Gdańsk–Sopot (2003)
59. Lewicki, A.: Rola abstrakcji pozytywnej i negatywnej w procesie uczenia się pojęć. Studia Psychologiczne **3**, 5–47 (1960)
60. Strawson, G. Mental Reality. MIT Press, Cambridge (2009)
61. Fodor, J. A. LOT 2: The language of thought revisited. Clarendon Press: New York (2008)
62. Thagard P. Mind-Society: From Brains to Social Sciences and Professions. Oxford University Press: Oxford (2019)

Chapter 2
The Role of the Concept of Mental Representation

Daniel Dennett's concept of the intentional stance, interpretivism, is unique among the theories discussed in this work as it is not an account of mental representations, but of intentionality. Intentionality is an essential feature of mental representations, though it is not exclusive to systems endowed with mental representations. Therefore, the class of intentional systems is broader than the class of systems endowed with mental representations. Dennett's defended concept of intentionality thus allows for the identification of the broadest class of cognitive systems that can be attributed mental representations. However, Dennett's intentional stance, as the only account discussed in this book, does not address the question of what the content of mental representation is or how it acquires reference.

Although Dennett's intentional stance, through the concept of intentionality, allows us to define the subject of my inquiry, which is mental representation, it does not in itself require the existence of real mental representations as separate entities in the cognitive system (especially at lower levels of system organization). It does not determine much about the role of mental representations in cognitive systems, requiring only that representations serve to explain and predict behaviors. Therefore, Dennett's characterization of intentionality only partially defends mental representations against anti-representationalism. It emphasizes the importance of description and explanations of behavior in intentional terms, which also characterize mental representations. However, without providing a concept of content or reference, it does not exclude instrumentalism regarding mental representations. Only with an adequate concept of content and reference can one defend the significance of representations in the system, and thus fully justify realism about mental representations.

In Sect. 2.2, I outline the relationship between intentionality and mental representations. I show that Dennett's concept of the intentional strategy is consistent with realism, as Dennett considers only those descriptions and explanations that achieve predictive and explanatory success to be intentional.

K. Bielecka, *What is Misrepresentation?*, Studies in Applied Philosophy, Epistemology and Rational Ethics 71, https://doi.org/10.1007/978-3-031-84375-4_2

In Sect. 2.3, I further characterize the concept of the intentional strategy. I demonstrate what can be described and explained in intentional terms. Among the three stances (intentional, functional, and physical), I focus on the first, which describes intentional systems. I show that in intentional terms according to Dennett, one can describe, explain, and predict human and animal behavior.

In Sect. 2.4, I outline the relationship between intentionality and the rationality of actions. To this end, I introduce Dennett and Davidson's principle of rationality, as well as refer to Kmita and Nowak's idealizational principle. I demonstrate that the rationality of actions is a necessary condition for the possibility of attributing mental representations. I emphasize the idealizational character of the rationality principle and its heuristic significance, which means that this principle only partially allows us to explicate how the adequacy of mental representation is connected with the correctness of actions.

In Sect. 2.5, I show that the intentional stance is applied to describe explanations and predictions of various rational behaviors of the cognitive system at different levels of its organization, both at the personal and subpersonal levels, in Dennett's sense.

In Sect. 2.6, I present some limitations of description and explanation in intentional terms. To this end, I invoke the criticism regarding the liberality of the intentional strategy (formulated by Ned Block). I turn it into an argument for the usefulness of the intentional stance. I show that the intentional strategy is used to describe rational systems but is insufficient for a complete psychological description in the functional sense (as would be required by the famous case of Phineas Gage).

I then provide a summary, taking into account the desiderata that emerge in the chapter concerning the concept of representation.

2.1 The Philosophical Concept of Intentionality

Intentionality is most generally understood as aboutness: it always directs us towards the object of representation, regardless of whether it is a representation of an existing or non-existing object.

Franz Brentano, in discussing mental phenomena, referred to the then somewhat forgotten concept of intentionality. In his theory, intentionality (i.e., aboutness) means that thoughts (in more contemporary terminology: beliefs and desires) are about something:

> Every mental phenomenon is characterized by what the Scholastics of the Middle Ages called the intentional (or mental) inexistence of an object, and what we might call, though not wholly unambiguously, reference to a content, direction toward an object (which is not to be understood here as meaning a thing), or immanent objectivity. (…) This intentional in-existence is characteristic exclusively of mental phenomena. [1: 68]

Brentano distinguishes between the object (the entity towards which the mental phenomenon is directed) and the content of representation, which were also written about by Kazimierz Twardowski [2] and Gottlob Frege [3].[1]

Dennett's Interpretivism remains consistent with the most rudimentary formulation of intentionality found in Brentano. However, a full explication of the concept of intentionality requires the concept of representational content. Dennett's account explicates this only partially, limiting itself to describing the role of intentional explanations in everyday thinking, philosophy, and science. By justifying the significance of the concept of intentionality, interpretivism paves the way for the analysis of the concept of the content of mental representations in subsequent chapters.

Brentano, as seen in the above quote, treats intentionality as a criterion for distinguishing mental phenomena from physical phenomena. Physicalism asserts that everything that exists is physical. According to naturalists, who typically embrace physicalism, mental phenomena are also physical. Thus, naturalists must reject Brentano's dualism while simultaneously determining why certain, but not all, physical systems are characterized by intentionality. This is the fundamental task of naturalized semantics [5, 6].

2.1.1 The Philosophical Concept of Beliefs and Desires and the Pluralism of Mental Representations

In twentieth century analytic philosophy, folk psychology was most often understood in terms of beliefs and desires. The philosophical concept of belief and desire is characterized in terms of propositional attitudes, that is, as someone's attitude towards a proposition. The basic form of such attributions of mental states to intentional systems are sentences that express so-called propositional attitudes.

x is convinced that p
y wants q to happen
z fears that r

Such sentences consist of three parts: a term referring to the intentional system in question (x, y, z), a term for the attitude attributed to it (belief, desire, wonder, . . .), and a term for

[1] An explication of intentionality would require explaining how it happens that intentionality "is directed at" some object through some content. This is precisely what Edmund Husserl [4] and his contemporary successors addressed. Husserl realized that such an understanding of intentionality implies the problematic assumption that intentionality must always have an object. This raises the question of how to explain the fact that we refer intentionally to non-existent objects (e.g., Papa Smurf) or merely possible or abstract ones (e.g., my imagination of myself in 50 years, the number π), and moreover, how this way of referring differs from intentionally referring to existing objects. These are certainly important and interesting studies that can be found in the works of Husserl and other phenomenologists. Due to the limited scope of this work, I will set these considerations aside. However, it is worth noting that various attempts to solve Brentanian problems through naturalistic concepts lead to well-known difficulties (see, for example, the account of Fred Dretske, which I will analyze in Chap. 7).

> the particular content or meaning of that attitude—the proposition denoted in these dummy cases by the letters *p*, *q*, and *r*. [7: 45]

Propositions are the contents of propositional attitudes. In analytic philosophy, two fundamental conceptions of propositions are distinguished: the causal-pragmatic and the linguistic (this is not a dichotomy, but rather a typology). According to the causal-pragmatic conception, propositions are functions from the set of possible worlds to the set of logical values, and sometimes selected subsets of some set of possible worlds [8]. On the other hand, according to the linguistic conception, propositions are derived from the content of internalized speech acts. In this work, I do not resolve in favor of either conception, but will use a more general formulation, according to which a proposition is a non-linguistic abstract object that represents the truth conditions of a sentence:

> Every sentence, in the logical sense, remains in a semantic relation of expressing to the so-called "proposition," a non-linguistic abstract object that represents, in some way, the truth conditions of that sentence. [8: 87–88]

As such, the philosophical concept of belief and desire is weaker than the colloquial one: someone's possession of a belief reduces to their recognition of some proposition as true, while harboring a desire reduces to the wish that a proposition be true.

Dennett's intentional strategy encompasses a broad class of mental representations. It applies to all kinds of mental representations in systems to which the attribution of intentionality provides new explanations and predictions.

Bence Nanay argues that an entire group of semi-conscious or unconscious activities, typically referred to as examples of know-how, need not be captured within the framework of a complex network of propositional attitudes, especially when there are better, alternative ways of describing them [9]. An example of a simple activity that does not require description in terms of propositional attitudes is playing table tennis. Describing the automated movements of the paddle in terms of rapidly changing attitudes towards propositions about the ball's position would be an unnecessary complication, especially since there is a better alternative description in terms of lower-level sensorimotor representations. According to Nanay, describing motor activities in terms of propositional attitudes would not yield any explanatory benefits but would merely constitute a redescription.

Numerous examples of this kind serve as empirical evidence for rejecting the assumption that a theory of mental representations should describe representations exclusively in terms of beliefs and desires. However, this does not argue for the complete rejection of beliefs and desires. In this work, I argue for a weaker, pluralistic theory of mental representations that rejects the assumption that beliefs and desires are the only kinds of representations.

2.2 Significance of the Concept of Intentionality

In this section, I outline Dennett's concept of interpretivism, which encompasses the general concept of an intentional strategy or intentional stance. I demonstrate the role that intentional explanations play in everyday thinking, philosophy, and science. I will examine the concept of stances, including the intentional stance, more closely in Sect. 2.3.

Daniel Dennett, alongside Donald Davidson, is considered one of the two most important proponents of interpretivism [10]. I discuss only a part of Dennett's theory of representation, namely interpretivism, thus Dennett's account of intentionality.[2]

Dennett [12] realistically treats regularities in behavior, which he recognizes as real patterns. He remains a realist regarding beliefs and desires at the personal level, concerning the entire cognitive system, but he may be an instrumentalist regarding beliefs and desires at the subpersonal level, concerning parts of the cognitive system. I write more about the personal and subpersonal levels in Sect. 2.5. Dennett allows for the possibility that beliefs and desires at the subpersonal level correspond to specific inscriptions, such as a sentence in the language of thought: "The grass is green"; however, he considers the possibility that all beliefs and desires correspond to concrete inscriptions to be highly improbable [13]. Thus, he critically addresses Fodor's syntactic theory of representation [14].[3]

He also argues against Ryle's a priori anti-representationalism:

> My goal in this essay has been simply to explore—and perhaps improve our view of—some open, empirically researchable territory. Ryle said, aprioristically, that we *couldn't* be mental-representation manipulators; Fodor and others have said, aprioristically, that we *must* be. Some of the details of the computer metaphor suggest what we *might* be, and in so doing may eventually shed some light on what we are. [11: 225]

He also writes:

> (...) while I think that the computer metaphor, properly used, can liberate the theorist from Ryle's worries, it is often abused by ideologues of cognitive science. It is obvious, let us grant, that computers somehow represent things by using internal representations of those things. So it is possible (somehow) that brains represent things by using internal representations of those things. But if we are to move beyond that modest bit of metaphysical elbow room and actually understand how the brain might represent by analogy with computer representation, we had best be clear about just how it is computers represent. There are several ways, or styles, of computer representation, and only some of them are plausible models for ways, or styles, of mental representation in the brain. [11: 215]

Dennett acknowledges the reality of entities described by attitudes that are marked by their predictive and explanatory success. However, he treats certain entities instrumentally, which have not been fully legitimized and about which we can only infer their existence unreliably. Another example of the instrumental use of the intentional stance is the intentional description of evolutionary processes (the beliefs of Mother

---

[2] Distinct from interpretivism is his theory of mental representations, in which representation occurs at the functional level [11].

[3] I discuss Fodor's syntactic conception of representation in Chap. 3.

Nature); beliefs attributed to Mother Nature at the psychological level should be treated solely instrumentally, as they are not psychologically real.[4]

Dennett's interpretivism allows us to describe what a cognitive system is capable of achieving with mental representations. Thus, it describes its competence.[5] However, it does not allow us to describe how this is possible, that is, the performance or implementation of such competence—although this is precisely the subject of interest in cognitive science, psychology, or cognitive neuroscience. A description at the competence level is insufficient for a full characterization of mental representation. As Dennett asserts:

> The fact about competence models that provokes my "instrumentalism" is that the decomposition of one's competence model into parts, phases, states, steps, or whatever need shed no light at all on the decomposition of actual mechanical parts, phases, states, or steps of the system being modeled—even when the competence model is, as a competence model, excellent. [11: 75–76]

Dennett's model of the cognitive system's competence resembles the computational level in Marr's sense. David Marr distinguished three levels of explanation for computational systems: the computational level, the representational level, and the implementation level. Descriptions at the computational level concern the functioning of the system in its environment and why this functioning is adequate in relation to the environment; descriptions at the representational level concern the algorithms and representations underlying the functioning of the system; descriptions at the implementation level concern how the system is physically realized [16]. Dennett treats the description occurring at the competence level (at the computational level, as understood by Marr) as an idealization, which, like any idealization, has its limitations. The description at the competence level does not have to correspond exactly to what happens at the implementation level (at the algorithmic level in Marr's sense), and it is not even advisable that it should correspond so.

One must always consider the level of implementation, which limits the idealized description "from below," showing its boundaries. However, this does not mean that either description—at the level of competence or implementation—sets the ultimate criterion for the correctness of the description of a phenomenon. The role of both levels is somewhat different: the level of competence defines the subject of study and in this sense selects from the level of implementation what is related to the subject of the description, while the level of implementation sets the boundary of the description from the level of competence, as the dependencies occurring at the level of implementation are real. In this sense, complete arbitrariness of the idealized

[4] The concept of the intentional stance is akin to the notions of the personal stance and the Darwinian stance, highlighting that rationality is not exclusive to persons but also extends to animals and even evolutionary processes (Mother Nature). The Darwinian stance is the only one used metaphorically, as Mother Nature is characterized by quasi-rationality. Dennett attributes to her what he calls free-floating rationales [15].

[5] I have borrowed the distinction between competence and performance levels from Noam Chomsky.

description is excluded by the real dependencies that exist in the world, just as complete fidelity to the description at the level of implementation is not possible.[6]

In summary: the concept of intentional strategy is consistent with realism regarding mental representation. The application of the intentional strategy only makes sense when it yields explanatory and predictive benefits. For a realist, the existence of such benefits is an argument for the existence of mental representations. Dennett does not have to deny their existence, as he treats instrumentally only those entities for which there is no clear empirical evidence.

2.3 Intentional Stance

Dennett characterizes intentionality using the concept of an intentional system. By "system," the philosopher understands a complex object containing many parts, among which interactions can occur. The behavior of such systems is governed by specific regularities, which we learn about through methods specific to the type of system. For example, a computer system consists of electronic parts and software; it operates according to the rules of the software (and the laws of physics). Knowledge of how a system, in this case a computer system, works can be drawn from observations, reading technical books, conversations with specialists, etc.

An intentional system is characterized by what Dennett calls the "intentional stance.“ This is not the only kind of stance—in Dennett's view, there are three fundamental types of stances, or three different approaches to explaining and predicting: physical, design, and intentional stance. I will briefly explain what each of the three stances entails and what distinguishes the intentional stance.

It is important to note initially that the adequacy of stances does not necessarily depend on strict laws or law-like regularities. The exception is the physical stance, which relies on such laws and law-like regularities. The stances themselves, that is, the strategies of description, are not the hard nomological-deductive rules or law-like regularities characteristic of scientific theories, though they are useful for explanatory purposes both in everyday practice and in scientific theories. The adequacy of stances depends on their explanatory and predictive success. Adopting a stance commits one to accepting more or less restrictive rules for explanation and prediction, depending on the stance taken. However, this does not mean that adopting a stance is irrevocable in a specific case. In the application to scientific theories, stances are subject to empirical verification by us and are, in a given case, accepted or rejected based on that.

The physical stance views reality as physical and employs the methods of the physical sciences. For example, from the physical stance, one can study interactions between atoms or celestial bodies. The physical stance always works because physical laws are universal and exceptionless (although they may be probabilistic). It is

[6] For more on the relationship between the competence level and the performance/implementation level, see the beginning of Chap. 3.

utilized not only by scientists but also by ordinary people performing daily activities (Dennett believes that there exists commonsense physics). Sometimes the physical stance is inaccessible in practice (we do not always have a thermometer at hand to measure our temperature).

At times, a slightly more general design stance is better suited for explanation and prediction, although it is not always easy to distinguish it from the physical stance. The subject of the design stance is the functional structure described at a level of abstraction sufficient for predicting or explaining the system's behavior. For example, the column of mercury in a thermometer indicates the ambient temperature; even without knowing the physical properties of mercury, one can predict and explain how the thermometer will behave in response to temperature changes. The accuracy of predictions or explanations is limited by the possibility of physical damage (for instance, in the event of physical damage to the thermometer, its operation cannot be attributed to a design error).

The most abstract is the intentional stance, describing a rationally acting system or agent with beliefs and desires, attributed based on its place in the world and purpose of action. The description from the intentional stance is characterized by the least degree of detail among all three. The intentional strategy ensures the credibility of the description of the intentional system:

> *all there* is to being a true believer is being a system whose behavior is reliably predictable via the intentional strategy and hence *all there is* to really and truly believing that *p* (for any proposition *p*) is being an intentional system for which *p* occurs as a belief in the best (most predictive) interpretation. [17: 72]

The intentional stance is governed by commonsense principles and assumptions, various commonplace truisms, which explain the behaviors of others and oneself. Among them is the rule of attributing beliefs and the rule of attributing desires. The rule of attributing beliefs states that most beliefs are true; we learn truths, among other ways, through the senses, and becoming acquainted with something normally implies knowledge. The rule of attributing desires is derivative of the first. According to this rule, we attribute to the system desires for things that—according to its beliefs—are good for it.

The intentional strategy does not deny objectivity to mental states, and adopting it is necessary to defend the concept of belief in a naturalistic way. Based on the explanatory and predictive success in explaining rational actions that the intentional strategy brings, one should recognize intentionality as an objective property of beliefs and desires, as well as other types of mental representations. Dennett emphasizes the explanatory and predictive power of the intentional strategy.

Predictive success is not tied to the perfect accuracy of the intentional strategy:

> This inability to predict fine-grained descriptions of actions, looked at another way, is a source of strength for the intentional strategy, for it is this neutrality with regard to details of implementation that permits one to exploit the intentional strategy in complex case, for instance, in *chaining predictions*. [17: 68]

Unlike philosophers who argue that representations serve only for thinking, such as Jerry Fodor [18], Dennett assumes that representations also serve for acting.

Consequently, Dennett's concept of intentionality provides partial justification for the desideratum:

I.1. An adequate theory of representation takes into account their role in describing, explaining, predicting, and controlling behaviors.

It is important to note that Dennett does not postulate the causal power of representations. As a result, he does not fulfill part of desideratum I.1, which concerns intervening in behavior using mental representations. The requirement of the causal power of representations would exclude the instrumentalist treatment of some mental representations, such as representations of Mother Nature.

Moreover, the intentional stance, it should be noted, is not limited to the description of human behaviors. It is also used to describe the behaviors of other living organisms and even artifacts:

> To adopt the intentional stance toward these monkeys is to decide tentatively, of course—to attempt to characterize, predict, and explain their behavior by using intentional idioms, such as "believes" and "wants," a practice that assumes or presupposes the rationality of the vervets. A vervet monkey is, we will say, an intentional system, a thing whose behavior is predictable by attributing beliefs and desires (and, of course, rationality) to it. [11: 242]

Consequently, Dennett's concept allows us to justify the desideratum:

I.2. An adequate theory of representation concerns both linguistic and non-linguistic phenomena (e.g., perceptual) of humans and animals.

The view that animals also have mental states is not uncontroversial. The most well-known opponent of attributing mental states to animals is Donald Davidson, who in his "Rational Animals" points out that the only available empirical evidence for hypotheses concerning the content of beliefs is verbal communication. Since animals do not possess our language and cannot be asked what they think, verifying the content of their thoughts is impossible. Davidson argues as follows:

> Creatures with propositional attitudes and creatures without such attitudes are alike in that both can be conditioned to respond differentially to many of the same properties, objects, or types of event. This misleads us into thinking similar processes are going on in the brains of both sorts of creature. And of course there is much that is similar. But a creature with propositional attitudes is equipped to fit a new concept into a complex scheme in which concepts have logical and other relations to one another. Speechless creatures lack the conceptual framework which supports propositional attitudes. [19: 137]

Davidson thus presents the argument from holism. Holism asserts that one cannot possess a concept in isolation from the entire complex network of other concepts. The argument boils down to the claim that without language, it is impossible to have such a network. To substantiate this, Davidson would need to demonstrate that assuming the existence of a conceptual network without language leads to a logical contradiction. However, since he does not do this, his argumentation appears to be unsubstantiated. Davidson is correct in stating that language is a sufficient condition for having a complex, holistic network of concepts. Yet, this does not imply that it is necessary.

Elsewhere, Davidson writes:

> *First*, I argue that in order to have a belief, it is necessary to have the concept of belief.
>
> *Secondly*, I argue that in order to have the concept of belief one must have language. [20: 324]

Davidson's argument has two steps. The first is:

D1) To have a belief, one must have the concept of belief.

The second states:

D2) To have the concept of belief, one must have a language [20].

Davidson does not explicitly refer to the concept of mental representation. Assuming that the concept of belief is a kind of mental representation, I propose to reformulate D1 as the thesis:

D1*) To have a belief as a particular kind of mental representation, one must have the concept of belief.

D1* expresses the condition that having concepts is to have second-order representations, that is, having representations about representations [21]. According to Davidson, having second-order representations requires having a language. Although animals have discriminative abilities, such as distinguishing between different species of animals and humans, they do not have the concept of belief because they do not have a language. These discriminative abilities are not equivalent to having beliefs, as the exact content of the supposed animal beliefs is unknown. If a dog wags its tail at a bone, it is unclear whether it represents the bone as a bone, food, retinal stimulation of its eye, etc. Since there is no way to communicate with the dog that would allow determining the reference of its representations, it is impossible to precisely establish such a reference.[7] Davidson, in *The Problem of Objectivity*, provides another example, namely the surprise in a worm, arguing that it is impossible to say that the worm is surprised and by what ([22]). Therefore, a necessary condition for having beliefs with a specific content is the possession of language, as only through language can one precisely attribute content to beliefs by speaking about them. A condition for the possibility of determining content is the ability to refer to individual beliefs, which in turn requires the concept of belief.

Contrary to Davidson, the problem of content determination can be resolved without language. This is pointed out by Tyler Burge [21], who recognizes that the mechanisms of perception in animals help define what their representations refer to. A dog, regardless of its distance from a bone, responds to it (behaviorally and neurologically) as if it were the same size; thus, the dog exhibits perceptual size constancy. The mechanism of perceptual constancy is not for detecting retinal stimulations, which allows us to exclude the variable retinal stimulations from the content of the dog's beliefs. In the case of a worm, there is no such constancy, so it is not possible to say that the object of surprise could be, for example, the appearance of Davidson's face. In the teleosemantic part of the book, I show that perceptual mechanisms, such as the mechanism of size constancy, are insufficient to fully resolve the problem of content determination, but there are methods that allow for resolution

[7] The problem highlighted by Davidson is known as the disjunction problem, which I discuss further in the chapter on causal theories and in the second part of the book.

without recourse to language, relying instead on evolutionary factors and cognitive architecture.

In summary: to solve the problem of content determination, one does not need to resort to communication, therefore Davidson's claim D1* is unwarranted.

Furthermore, the premise D1* can be challenged by pointing out that numerous non-linguistic animal behaviors can be described as practical inferences aimed at goals such as capturing prey or avoiding predators. There is empirical evidence suggesting that animals utilize episodic memory and cues present in the environment. These findings support the hypothesis that such behaviors are practical inferences and, moreover, can be considered representational [23].

I reject the premise D1 (and D1*) for these two reasons. First, it is possible to non-communicatively determine the content of ascribed beliefs; second, practical reasoning is commonly attributed to animals, describing their behaviors. Therefore, I conclude that Davidson's argument is valid but unsound, as it contains a false premise. For this reason, I also reject his argument that rationality requires communication and that it is not possible to attribute beliefs to animals.

Let us return to the applications of the intentional stance. It is also useful when the research question is normative or hermeneutic. Thus, it serves not only to explain but also to critique or praise the actions and intentions of individual agents or entire communities. In such cases, it abstracts from the psychological states of subjects. For instance, when interpreting the provisions of a law, we do not examine who specifically passed it or proposed amendments to it. In interpreting the statutes, we refer to an idealized rational agent, such as a perfectly rational legislator, who avoids contradictions with the constitution or other legal acts, etc. Similarly, a literary critic, analyzing a plot, most often abstracts from the biography of the author, for example, of a novel; they also leave aside the motivations and psychological states of the author (with the exception, of course, of autobiographical books).

The application of the intentional stance is thus broader, as it extends beyond psychology and broadly understood cognitive science.

Within Dennett's interpretivism, he does not provide concepts of content and reference. This is not a complete characterization of the concept of mental representation. However, the intentional strategy facilitates the determination of the cognitive architecture of the organism, which is not merely an empty "black box":

> The intentional stance, however, provides just the right interface between specialties: a "black box" characterization of behavioral and cognitive competences observable in the field, but couched in language that (ideally) heavily constrains the design of machinery to put in the black box. [15: 350]

Nevertheless, the fact that the intentional strategy facilitates the determination of cognitive architecture, that is, its functional structure, does not mean that the intentional strategy replaces the design stance. The intentional strategy—as Dennett emphasizes—"presupposes and does not explain rationality or intelligence" [24: 99]. In other words: by attributing, for example, desires and beliefs, we do not resolve what their psychological nature or functional substrate is. Since it does not resolve what the psychological nature or functional substrate of representations is, no theses on this subject can be legitimized by the predictive or explanatory success of the intentional stance. To justify such claims, one must use the design stance. I write more about the limitations of the intentional strategy in Sect. 2.5.

Anti-representationalists challenge the notion that the attribution of representations can lead to accurate predictions and explanations of cognitive processes, an assumption central to Dennett's argument for the intentional stance. Dennett counters the eliminativists' critique of mental representation by demonstrating the utility of employing intentional concepts in both everyday and scientific practice. To illustrate this, he introduces the figure of a Martian [17]. Martians, unlike Earthlings, do not use folk psychology to explain the behaviors of others, but they do know all the physical facts about a person:

> Take a particular instance in which the Martians observe a stockbroker deciding to place an order for 500 shares of General Motors. They predict the exact motions of his fingers as he dials the phone and the exact vibrations of his vocal cords as he intones his order. [17: 69]

For a Martian, who does not employ folk psychology, predicting the behavior of a person using everyday explanations would seem miraculous. There is no method of fact selection within the physicalist approach, as there is no method of fact generalization concerning human behaviors from all physical facts. Eliminativists regarding folk concepts (such as Paul and Patricia Churchland) would have to adopt folk psychology to have these generalizations. Anti-representationalists who do not deny folk psychology would have to accept that representations play a role in explaining and predicting cognitive processes. However, this would make them closet representationalists. Therefore, they must reject Dennett's thesis and maintain that there exists a non-intentional but equally general method of fact generalization about behavior.

Many anti-representationalists turn to dynamic systems theory as sufficient for describing cognitive processes [25, 26]. They would argue that Martians, using physical methods, could identify parameters of a dynamic system that would suffice to predict behavior in accordance with the principles of simplicity and parsimony. In other words, they reject the assumption that such a method is unavailable to Martians. This assumption, they claim, begs the question: Dennett's error is thus a case of petitio principii. In defense of Dennett, however, it can be said that to date, dynamicists have not demonstrated such a universal method of explanation and prediction, and their claim to its universality remains, for now, an unsupported promise [27].

The characterization of various stances leads to the conclusion that the concept of representation, if it is to be useful in explaining and predicting cognitive processes, cannot be described solely by one of the three stances. To the extent that Dennett would agree with anti-representationalists, it is in recognizing that the concept of mental representation defined solely by the intentional stance does not capture the essence of representation, saying relatively little—for example—about its functioning at the level of organism biology. Describing the functioning of intentional concepts in practice is insufficient for fully characterizing the realization of representations. On the other hand, limiting the description of the concept of representation to a physical strategy, or even a physical and functional one, does not suffice for a complete explanation of mental representations, as it remains unclear what is being explained. A functional description is limited to the description of the mechanism of action, and therefore the functional characterization of this mechanism, without indicating the subject of explanation and prediction, is explanatorily inefficient for explaining and predicting representations. I write more on this subject in the following chapters.

2.4 Intentional Strategy and Rationality

The intentional strategy describes the behaviors of cognitive systems characterized by instrumental rationality. The actions of rationally instrumental cognitive systems can be successful when they select the appropriate means for a previously intended goal. Mental representations can cause these actions: correct representations usually lead to successful actions, while erroneous representations typically lead to failures in action. The task of a naturalistic theory of mental representations is to explicate the relationship between mental representation and the accuracy of actions in such a way that it is psychologically adequate.

The rationality of actions is explained by the principle of rationality or charity [28, 29]. The principle of rationality dictates that we perceive cognitive agents as subjects having true beliefs, which commonly share certain basic beliefs about the world. According to the principle of rationality, we attribute to others true beliefs that are coherently interconnected, and we assume that they do not contradict each other:

> If we cannot find a way to interpret the utterances and other behaviour of a creature as revealing a set of beliefs largely consistent and true by our own standards, we have no reason to count that creature as rational, as having beliefs, or as saying anything. [28: 324]

Davidson further emphasizes the importance of the principle of rationality for communication. He asserts that the assumption of the consistency of the set of beliefs of each participant is a condition for successful communication:

> ... just as we must maximize agreement, or risk not making sense of what the alien is talking about, so we must maximize the self-consistency we attribute to him, on pain of not understanding *him*. [30: 313]

The principle of rationality is an idealizing principle. This means that it obliges the interpreter to regard someone's beliefs as true and coherently related, as long as this is possible in light of the interpreter's knowledge and epistemic goals, as well as in the given context of the statement [30, 31].[8] The idealizing principle of rationality limits the set of possible attributions of content to the cognitive system's representations in such a way as to preserve its rationality. As Davidson (and Dennett

---

[8] Davidson emphasizes that many shared beliefs are practically imperceptible: "If we fail to notice the extensive range of mutual agreement and shared assumptions during communication, it is only because there are too many truths that we share, and they are too obvious to attract our attention. What we aim to discuss are matters that are new, strange, or debatable" [32: 321].

too) emphasizes, attributing rationality in an idealized way enables understanding why a cognitive system can be irrational. A common agreement on certain things forms the basis for understanding what irrational beliefs are:

> Making sense of the utterances and behavior of others, even their most aberrant behavior, requires us to find a great deal of reason and truth in them. To see too much unreason on the part of others is simply to undermine our ability to understand what it is they are so unreasonable about. [32: 321]

I will illustrate the last sentence from the quote with the example of John (from the introduction). Imagine Johnny, who wants to take the bus to school. John remembers an outdated bus schedule. He does not know that according to the current schedule, the bus no longer arrives at that time, and therefore he still waits for the bus. Until we learn that John has the wrong schedule, we will be inclined to attribute to him a true belief about the bus's arrival. We are inclined to attribute to John, standing at the bus stop, a whole set of various true beliefs, concerning, for example, that buses stop at bus stops, that buses do not consume people, that the sun rises every day.

The principle of rationality also plays the role of a heuristic in explaining the behaviors of humans and animals. I understand the concept of heuristic as follows:

> An algorithm (...) is a set of rules which, if followed literally, will achieve the desired solution. A heuristic, on the other hand, is a set of rules which, if followed, may achieve a solution but cannot guarantee doing so. [33: 83]

Thus, the Principle of Rationality is a set of rules whose application usually leads to the correct attribution of intentional states, but it does not necessarily have to do so.

In summary: the Principle of Rationality is an idealizing principle, as it sometimes deliberately distorts the description of cognitive system behavior in such a way as to maintain the truth and coherence of its actions (depending on contextual factors, such as the interpreter's knowledge). At the same time, it has heuristic roles, as it usually allows for the correct attribution of intentional states, making it useful in communication.

The Principle of Rationality is partly related to the fallibility of mental representations. As an idealization, the Principle of Rationality allows us to set aside systematic error, thereby making it possible to preserve rationality and explain irrational behaviors. However, being merely a heuristic, the Principle of Rationality does not allow for a precise determination of when we can consider mental representations to be erroneous or correct. The Principle of Rationality does not provide a positive concept of representational error, which would allow us to determine when the behaviors of the cognitive system are irrational. Therefore, it requires supplementation with a concept of representational error, without which a full explication of the relationship between mental representation and rationality is not possible. Dennett's conception, which utilizes the Principle of Rationality, only shows a partial relationship between mental representation and rationality and thus partially fulfills the Desideratum of Rationality:

II.1. An adequate concept of representation is consistent with the assumptions regarding the rationality of the cognitive system.

I write more about the relationship between representation and rationality and the Desideratum II.1 in the teleosemantic section.

I will now illustrate some limitations that the Principle of Charity, as an idealizing principle, brings with it when taken literally. Its literal formulation is in the form of a maxim:

M1) Interpret people as irrational only when you have overwhelming evidence to do so [34].

I refer to the example from Jerzy Kosiński's well-known story "Being There" [35]. The author describes the life of a simple illiterate man, the gardener named Chance, who spent his entire life tending to the Old Man's garden. Chance had no other contact with reality apart from television, which he very much enjoyed watching. His worldview was a vision from Hollywood movies, where everything was neatly arranged and, despite the twists and turns, always ended well. When the Old Man died, Chance found himself alone in the world and, by accident, met the wife of a wealthy businessman who, having misheard his real name, mistook him—or perhaps did so intentionally, hearing what she wanted to hear—for Chauncey Gardiner, a person from a wealthy entrepreneurial family. From then on, Chance, a man without a past, acquired an identity and adapted to it remarkably well, having learned that one acquires characteristics from the outside, just as observers do with television characters. Ross did not play his role but answered others' questions honestly and without beating around the bush, (unwittingly) allowing others to interpret his statements. All off-topic remarks, when Ross, out of politeness, spoke about things he had the faintest idea about, were taken as profound, metaphorical, or witty remarks, and Ross himself was regarded as an important figure. Ross's response to a question about the state of the current economic policy, though completely irrelevant and like the vast majority of his answers, related to gardening, gained fame in such circumstances as a fitting metaphor for the American economy. To the other characters, coming from the upper and middle class, Ross appeared as an exceptionally competent person, and what's more—noble and tactful. The naive and often senseless statements of the protagonist of "Being There" were interpreted by others, in accordance with the principle of rationality, as true and rational utterances. Paradoxically, a man limited and raised in isolation from most social conventions was kindly treated as a man well-versed in society.

The literary example demonstrates that the principle of rationality, in its literal articulation, results in misdirection. It is possible that we lack compelling evidence of someone's irrationality, and in such cases, we may treat irrational behaviors as rational. In the scenario described in the story, Ross's statements and behavior should not be treated lightly or dismissively as rational behavior, but rather should be taken literally as irrational. However, the characters in the story who acted in accordance with the principle of rationality rationalized the unwise beliefs of Gardiner: in their eyes, the main character's beliefs, taken literally as irrational, were deemed rational.

As an idealizing principle, the principle of rationality of Dennett and Davidson aligns with the assumption of rationality in the approach of Kmita and Nowak, as both abstract from the psychological aspects of the cognitive system [36, 37]. Kmita and Nowak emphasize that their assumption of rationality is typical for the humanities and thus remains independent of psychology. Naturalization of rationality is not required in the humanities. The intentional stance allows for independence from psychology. However, nothing prevents the intentional stance from being concurrent with a psychologically plausible explanation of cognitive systems, such an explanation can also be a research goal.

Kmita and Nowak idealize a different aspect of the rational behavior of the cognitive system. Dennett and Davidson's idealization concerns the coherence and veridicality of beliefs and desires, abstracting from their fuller characterization. In contrast, Kmita and Nowak's idealization pertains to the circumstances under which the intended goal is realized, as the assumption of rationality abstracts from the circumstances surrounding the behavior of the cognitive system; the agent's actions are taken under conditions of certainty. The assumption of rationality describes the actions of a rational agent who, based on the knowledge possessed at a given time t, will undertake an action A, invariably leading to the attainment of the desired outcome S.

The articulation of the assumption of rationality by Kmita and Nowak more pointedly shows that the principle of rationality, as an idealizing principle, maximizes the correctness of actions to exclude their fallibility. However, psychologically sound ideally rational actions do not exist, as all our actions are taken under conditions of uncertainty. Therefore, the assumption of rationality would require concretization to be both psychological and rational. Such concretization (or de-idealization) cannot be provided by the intentional description alone, which is not enriched by a credible functional account.

2.5 The Personal and Subpersonal Levels of Description

Dennett distinguishes between personal and subpersonal levels of description [38]. The description of the personal level pertains to the cognitive system as a whole within its environment, while the subpersonal level description concerns its parts.[9] Both levels, personal and subpersonal, can coexist.

Dennett characterizes explanations in psychology as "a mixed bag." Mixed bag explanations combine elements of folk psychology and scientific psychology, as well as intentional and functional explanations, personal and subpersonal levels. Dennett notes that different branches of psychology are governed by their own inherent logic, which adds to the difficulty (beyond the fact that the levels of explanation are interconnected).

[9] For more on the discussion regarding the interpretation of Dennett's distinction between the personal level and the sub-personal level, see [39].

Subpersonal level explanations allow us to understand, to some extent, the complex activities that sustain the functioning of the system as a whole. However, mechanical components can also be described using intentional labels, which can blur the boundary between a system from a functional stance and an intentional system:

> We can also take advantage of the intentional stance to explore models that break down large, sophisticated agents into organizations of simpler subsystems that are themselves intentional systems, subpersonal agents that are composed of teams of still simpler, 'stupider' agents, until we reach a level where the agents are 'so stupid that they can be replaced by a machine'—a level at which the residual competence can be accounted for directly at the design stance. [40: 344]

In most cases, the intentional stance is applied to describe what happens at the personal level (for example, a child's rude behavior at the table is explained by the child's unfamiliarity with the rules of etiquette). However, there are instances where functional explanations are found at the personal level, while intentional explanations occur at lower levels of description. Dennett allows for intentional description at both the personal and subpersonal levels, as well as functional description at the personal level. An example of applying the intentional stance at the subpersonal level is found in visual representations in David Marr's theory [16]. According to him, visual representations operating at the lower, subpersonal level contribute to vision at the higher, personal level. Representations from the lower level can be described intentionally, as they are about something, although their object is different from the object of vision of the whole subject: they are about the properties of seen objects, which become increasingly distal at each stage of the vision process. Marr's example of visual representations shows that the object of subpersonal intentional description can be different from the object of intentional description at the personal level. Meanwhile, the object of personal intentional description of representations of something warm may be the same as, for example, the object of intentional description of the thermoreception module.

The division into personal and subpersonal levels is consistent with so-called homuncular functionalism, which Dennett advocates [40]. The idea of homuncular functionalism is that there is no single central homunculus managing the functioning of the cognitive system, but rather that functioning occurring at various levels of organization is equally important. This allows Dennett to avoid the undesirable consequences of postulating a homunculus. Gilbert Ryle [41] argued that the existence of a homunculus is assumed by theories that try to explain intelligent actions by postulating other intelligent actions, which in turn require explanation by postulating yet more intelligent actions, leading to an infinite regress. In Dennett's framework, successive homunculi are progressively less intelligent.

Various strategies are applied differently to describe behavior at both the personal level and subpersonal level. The division into these two levels allows us to recognize a particular difficulty associated with identifying the functional substrate of intentional states. Do intentional states correspond to specific functional states in a one-to-one correspondence? Or perhaps some complex relationships between functional states strictly correspond to intentional states? Finding such a correspondence, that is, the localization of the substrate of the intentional state, is decidedly more challenging for systems with rich internal organization, where there are more diversely interconnected mental representations.

Since intentional concepts are applied at different levels, mental representations can also occur at various levels. Dennett's concept of stances is consistent with the desideratum:

> I.3. An adequate theory of representation concerns phenomena understood at both the personal level and the subpersonal level.

The following objection is encountered against postulating representations at the lower level. Jonathan Francis Bennett and Peter Hacker argue that postulating representations at the lower level is a mistake because the whole (i.e., the entire subject at the personal level) and its parts (at the subpersonal level) cannot possess the same property:

> Talk of the brain's perceiving, thinking, guessing or believing, or of one hemisphere of the brain's knowing things of which the other hemisphere is ignorant, is widespread among contemporary neuroscientists. This is sometimes defended as being no more than a trivial *façon de parler*. But that is quite mistaken. For the characteristic form of explanation in contemporary cognitive neuroscience consists in ascribing psychological attributes to the brain and its parts *in order to explain* the possession of psychological attributes and the exercise (and deficiencies in the exercise) of cognitive powers by human beings. [42: 7]

Indeed, parts do not always possess the same properties as the whole, but sometimes they do (if a table is monochromatic, then the legs of the table are of the same color as the rest of the table). Settling for such a statement would not be controversial. Therefore, the claim that the property of intentionality does not belong to the whole person in the same way as to its parts requires further justification. For philosophers of language like Bennett and Hacker, the impossibility of attributing intentionality to parts of a system is justified by the argument from ordinary language use. They believe that colloquial language prohibits attributing intentionality to the entire person in the same sense as to parts of the system (the brain or the nervous system). From the language argument, Bennett and Hacker infer the ontology of the mind we should adopt. Their premise is actually false, as the usage of language is different: people are increasingly using phrases such as "the brain interprets reality" or "one hemisphere conveys something to the other." Thus, usage is changing in favor of the opponents of this argument.

The authors of the argument would essentially have to appeal not so much to the use of language as to an additional assumption about its nature, such as the idea that colloquial language expresses the nature of the world (essentialism). Adopting essentialism, however, requires separate argumentation, independent of linguistic practice. Bennett and Hacker, however, have no other argument than reference to linguistic practice, and thus their argument is circular. The practice of language use in science shows that an essential feature of language is its variability, as noted by Bartosz Brożek, who points out that everyday and scientific practice remains in constant dynamics [43]. Some metaphorical uses (among them may be the phrase "the brain sees") can be rejected as metaphors that inaccurately describe reality. Treating language as having an essential nature could be considered a hidden assumption of Bennett and Hacker's argument—then, as Brożek and others argue, it would be the error of apriorism. It is precisely apriorism that Dennett accuses Bennett and Hacker of:

> If Hacker were able to *show us* the rules [of proper word usage – K.B.], and show us just how the new uses conflict with them, we might be in a position to agree or disagree with him, but he is just making this up. [42: 85]

Dennett thus calls into question the existence of rules that Bennett and Hacker refer to. Their argument, that this is an incorrect usage, would have to appeal to linguistic reasons. Meanwhile, in their critique of neuroscience, they smuggle in metaphysical assumptions about intentionality.

2.6 Limitations of the Intentional Strategy

In this section, I demonstrate that the intentional stance maintains partial independence from other descriptive strategies, the design stance and the physical stance. An intentional description can be complemented by a functional or physical description, as I show using the well-known medical history case of Phineas Gage's brain dysfunction. At the same time, a functional description does not mean abandoning the intentional description. I show—contrary to Block, the author of the thought experiment on an artificial cognitive system (called Blockhead)—that his example can be considered an argument supporting the application of the intentional strategy in the description and explanation of behaviors of rational systems.

Dennett emphasizes that a description in intentional terms makes sense when it yields a benefit in the form of accurate predictions that are new or valuable. He also shows that there are cases where applying the intentional stance is a misuse. The lecturer's podium is not an intentional system because adopting the intentional stance does not yield new or more valuable predictions about its behavior:

> For instance, it seems the lectern in this lecture room can be construed as an intentional system, fully rational, believing that it is currently located at the center of the civilized world (…), and desiring above all else to remain at that center. (…) I predict the lectern's behavior, accurately, from the intentional stance, so is it an intentional system? If it is, anything at all is. [17: 66]

However, without attributing desires to the podium, we know that podiums do not move on their own. In such a case, the intentional stance is unnecessary to explain why the podium remained in the same place.

The intentional strategy is sometimes supported by facts concerning the functioning of the cognitive system, thus by the design stance. In psychological explanations of behavior, there are cases where referring to the intentional strategy is insufficient to explain and predict someone's behavior, and then it is necessary to refer to the design stance. Phineas Gage suffered significant damage to his frontal lobes in an accident in which a steel rod pierced his brain. Gage experienced significant personality changes, which Antonio Damasio wrote about:

> The most striking aspect of this unpleasant story is the discrepancy between the normal personality structure that preceded the accident and the nefarious personality traits that surfaced there after and seem to have remained for the rest of Gage's life. Gage had once known all he needed to know about making choices conducive to his betterment. He had a sense of personal and social responsibility, reflected in the way he had secured advancement in his job, cared for the quality of his work (...). He was well adapted in terms of social convention and appears to have been ethical in his dealings. After the accident, he no longer showed respect for social convention; ethics in the broad sense of the term, were violated; the decisions he made did not take into account his best interest, and he was given to invent tales "without any foundation except in his fancy," in Harlow's words. There was no evidence of concern about his future, no sign of forethought [44: 10–11][10]

Gage's behavior before and after the accident differed greatly: from a polite, responsible, and organized individual, he became overexcited, vulgar, and incapable of consistent planning. As Damasio put it, "Gage lost something uniquely human, the ability to plan his future as a social being" [44: 19]. Gage's case falls into the broader category of patients with serious brain damage, whose behavior does not follow the same rules of interpretation as that of healthy individuals.

Gage's case could be treated as a challenge to the intentional strategy and simultaneously as an argument for abandoning the intentional description in favor of a functional one. On the one hand, it is impossible to create a description of Gage's behavior as a rational being in the ideal sense; his behavior cannot be explained or predicted because it is not governed by any detectable rules characteristic of the behavior of a rational person in the ideal sense. Whenever Gage attempts to make a decision, it can be predicted that he will not consistently adhere to the plan and, for example, instead of going to work, he may change his mind halfway there and go to a pub. Only a more detailed description of his behavior in functional terms, e.g., in neuronal terms, allows for a better understanding of his actions. Describing Gage's behavior using a functional strategy allows for a better explanation of why his behavior is not treated by us as rational in the ideal sense.

On the other hand, Gage's rudimentary behaviors can be described as rational to a lesser degree. Understanding the neuronal causes of Gage's actions does not preclude treating him as a rational being. He likes to drink, so he goes to the pub; that is rational for the immediate future. Empirical facts weaken the concept of perfect rationality, thus they are an argument in favor of adopting rationality in a more limited sense, since Gage is incapable of maintaining plans over the long term.

Gage's case, therefore, is not a serious objection to the use of the intentional stance. According to Dennett, different attitudes can be combined, and not all intentional systems need to be rational to the same extent.

Ned Block constructs the following thought experiment about a machine called "Blockhead." Instead of a brain, Blockhead has a lookup table that contains a vast, yet

[10] But see [45] for a critical reassessment of the credibility of Harlow's account, on which Damasio's account is heavily dependent. This does not change the overall point that neurological damage may undermine our ability to rationality explain someone's behavior.

finite number of syntactically and grammatically correct sentences needed to carry on a conversation in a finite amount of time [46].[11] Although Blockhead's conduct is based on a complex internal structure, Block argues that we would still intuitively agree that Blockhead does not think and lacks intentional states. Our intuitions forbid ascribing intentional states to a machine that automatically responds to every question posed (it is, among other things, incapable of learning). Block's experiment aims to show that the intentional strategy is overly permissive, as it allows Blockhead to be classified among rational systems.

However, Block's argument seems unconvincing. Block constructs a thought experiment in which he assumes that the intentional stance is disconnected from the functional basis. Indeed, we also apply the intentional stance to describe the behaviors of hypothetically rational systems, and in doing so, we abstract from their psychology. An example is a judge who interprets a statute in light of the intentions of an ideally rational lawgiver, who is neither any individual legislator nor even the parliament on the day the law was passed. However, prima facie, Blockhead is comparable to an actual cognitive system whose behaviors we interpret across different scenarios, often engaging with them.

Block argues against the intentional stance, claiming that it fails to justify theses concerning the psychological nature of the functional substrate of beliefs and desires. Block assumes that actual cognitive systems should be rational due to their structure, not merely in light of their behaviors. However, the intentional strategy, which primarily focuses on behaviors, requires that they exhibit certain regularities. Thus, the intentional strategy would deny the possession of intentional states to automatic cognitive systems whose behavior reveals a lack of such states, for example, when responses during a conversation are detached from context. An example of such behavior could be an extended conversation with Blockhead. Contrary to Block's view, the intentional strategy would not deny intentional states to cognitive systems on the grounds that they are endowed with a particular structure. Moreover, the intentional stance also requires a certain internal constitution of the cognitive system so that the system's behavior betrays that it thinks, and thus has intentional states [10]. As Dennett states, the intentional stance only then has predictive power:

> But once we turn our attention to the truly interesting and versatile intentional systems, we see that this apparently shallow and instrumentalistic criterion of belief puts a severe constraint on the internal constitution of a genuine believer, and thus yields a robust version of belief after all. [17: 72]

Block's argument seems unconvincing for further reasons. Blockhead must be equipped with a vast amount of data that forms a long list of responses to the questions posed to it. Block assumes that, as interlocutors of Blockhead, we can easily

[11] Previously, a similar example was constructed by Stanisław Lem [47]: "It is a Gramophone, enlarged to the size of the planet (or the Universe). It contains very many, say, one hundred trillion, recorded answers to all possible questions. And thus when we ask a question, the machine does not "understand" anything at all, it is only the form of the question, that is, the order of vibrations in our voice, that starts the transmitter which plays a record or tape containing a prerecorded answer" (p. 130).

realize that it is an automatic system and suspend the application of the intentional strategy. However, the list of responses is so long that in trying to understand and predict Blockhead's responses in non-intentional terms, we would have to behave like the Martians from Dennett's example, or at least know the relevant functional facts about the described system. One might think that it would be easier to assume that Blockhead is a rational system, applying generalizations available through the intentional strategy. If Blockhead is appropriately designed, it might turn out that using the intentional strategy allows us to understand its behaviors more quickly and effectively. Therefore, the example of Blockhead could be an argument from simplicity for the application of the intentional strategy.

Furthermore, Block does not address the issue of the purpose of engaging in conversation with Blockhead. It seems that Block would need to provide an example of a situation where one could not interpret Blockhead in the same way we would interpret a computer playing chess [12]. Dennett argues that instead of understanding the highly complex structure of a chess-playing program, it suffices to assume that it operates according to the rules of the game; then one can play against it without almost any knowledge of the program's structure. However, one could then reiterate, following Dennett, the argument for the intentional stance and its predictive power.

In summary: Block is correct that a more comprehensive description of the system's behavior requires additional facts about its functioning if the subject of explanation is an actual cognitive system. However, Block is mistaken about the role of the intentional stance. Contrary to his view, the intentional stance does not serve to justify specific theses about the functioning of the cognitive system or its internal structure. Block's objection can be reversed and treated as an argument in favor of applying the intentional stance, not against it. It confirms the validity of adopting the intentional stance, which sometimes most simply allows for explaining and predicting intelligent behaviors without the need to postulate their functional underpinnings. Block's argument thus confirms Dennett's claim about the applicability of the intentional strategy for describing rational systems. It is not true that the intentional strategy is free from behavioral and structural assumptions about the cognitive system. Dennett emphasizes that this strategy assumes, rather than explains, the system's rationality. Contrary to Block, having a rich internal structure is a necessary condition for attributing intentional states to a cognitive system. The system's behavior is evidence of its sufficiently rich structure.

2.7 Summary

Dennett's concept of intentionality allows us to conceive of cognitive systems (including their organization) as rational. Representational systems are a subclass of intentional systems in Dennett's sense. Dennett's concept meets the desiderata: of universality (I.2), of multi-levelness (I.3), and of rationality (II.1), as well as partially the explanatory desideratum (I.1). Although it provides arguments for the significance of the concept of mental representation in everyday and scientific practice, it does not offer a concept of content and reference. Therefore, it does not defend the concept of mental representation against the criticisms of anti-representationalists, who primarily demand from representationalists an indication of what mental representation should do (thus a reference to the implementation level).

The concept of the intentional stance justifies why intentionality is important. It is so because it serves to describe, explain, and predict rational cognitive systems. Dennett believes that the intentional stance itself does not justify attributing to representations causal powers sufficient to change the behavior of intentional systems, that is, to intervene in behavior through representations. For this reason, it does not fully meet Desideratum I.1. The concept of the intentional stance does not explain what intentionality consists of, and this is a very important question for the naturalistic conception of mental representation.

In subsequent chapters, I seek both an adequate conception of content and reference, and a fuller explication of the concept of intentionality. I begin by analyzing concepts that locate the source of content in the internal structure of the cognitive system (Chap. 3). Then I examine concepts of content and reference that place reference-determining relations in interactions with the environment (Chaps. 4 and 5). Finally, I consider theories that attempt to account for the possibility of representational error (Fodor's asymmetric theory, teleosemantic theories in Part III).

References

1. Brentano, F.: Psychology from an Empirical Standpoint. Routledge, London (1874/1995)
2. Twardowski, K.: On the Content and Object of Presentations (Trans. Grossmann, R.M.). Nijhoff, The Hague (1894/1977)
3. Frege, G.: Über Sinn und Bedeutung. Zeitschrift für Philosophie und philosophische Kritik **100**, 25–50 (1892)
4. Husserl, E.: Logical Investigations (Trans. Findlay, J.N.). International Library of Philosophy. Routledge, London; New York (1970)
5. Haugeland, J.: The intentionality all-stars. Philos. Perspect. **4**(1990), 383–383 (1990). https://doi.org/10.2307/2214199
6. Schulte, P.: Mental Content. Cambridge University Press (2023)
7. Dennett, D.C.: Kinds of Minds. Basic Books, New York (1996)
8. Ciecierski, T.: Nastawienia sądzeniowe: wykłady z filozofii psychologii. Wydawnictwo Naukowe, Warszawa (2013)
9. Nanay, B.: Between Perception and Action. Oxford University Press, Oxford (2013)
10. Molder, B.: Mind ascribed: an elaboration and defence of interpretivism. John Benjamins Publishing Company, Amsterdam; Philadelphia (2010)
11. Dennett, D.C.: The Intentional Stance. MIT Press, Cambridge (1987)
12. Dennett, D.C.: Real patterns. J. Philos. **88**(1), 27–51 (1991)
13. Dennett, D.C.: Brainstorms. Philosophical Essays on Mind and Psychology. MIT Press, Cambridge (1978)
14. Fodor, J.A.: The Language of Thought. Thomas Y. Crowell Company, New York (1975)

15. Dennett, D.C.: Intentional systems in cognitive ethology: the "Panglossian paradigm" defended. Behav. Brain Sci. **6**(3), 343–390 (1983). https://doi.org/10.1017/S0140525X00016393
16. Marr, D.: Vision: a Computational Investigation into the Human Representation and Processing of Visual Information, 2nd edn. MIT Press, Cambridge (2010)
17. Dennett, D.C.: True believers: the intentional strategy and why it works. In: Heath, A.F. (ed.) Scientific Explanation: Papers Based on Herbert Spencer Lectures Given in the University of Oxford, pp. 150–167. Clarendon Press, Oxford (1981)
18. Fodor, J.A.: LOT 2: The Language of Thought Revisited. Clarendon Press, Oxford, New York (2008)
19. Davidson, D.: What thought requires. In: Problems of Rationality, pp. 135–151. Oxford University Press, Oxford (2004)
20. Davidson, D.: Rational animals. Dialectica **36**(4), 317–327 (1982). https://doi.org/10.1111/j.1746-8361.1982.tb01546.x
21. Burge, T.: Origins of Objectivity. Oxford (2010)
22. Davidson, D.: The problem of objectivity. In: Problems of Rationality, pp. 3–19. Oxford University Press, Oxford (2004)
23. Zawidzki, T.: Mindshaping: a New Framework for Understanding Human Social Cognition. MIT Press, Cambridge (2013)
24. Dennett, D.C.: Intentional systems. J. Philos. **68**(4), 87–106 (1971). https://doi.org/10.2307/2025382
25. Chemero, A.: Radical Embodied Cognitive Science. MIT Press, Cambridge (2009)
26. Garzon, F.C.: Towards a general theory of antirepresentationalism. Br. J. Philos. Sci. **59**(3), 259–292 (2008). https://doi.org/10.1093/bjps/axl007
27. Kirchhoff, M.D.: Anti-representationalism: not a well-founded theory of cognition. Res Cogitans **2**, 1–34 (2011)
28. Davidson, D.: Radical interpretation. Dialectica **27**(3/4), 313–328 (1973)
29. Davidson, D.: Inquiries into truth and interpretation. Clarendon Press, Oxford (1984)
30. Davidson, D.: Truth and meaning. Synthese **17**(1), 304–323 (1967). https://doi.org/10.1007/BF00485035
31. Davidson, D.: Toward a unified theory of meaning and action. Grazer Philosophische Studien **11**(1), 1–12 (1980). https://doi.org/10.1163/18756735-90000093
32. Davidson, D.: Belief and the basis of meaning. Synthese **27**(3), 309–323 (1974). https://doi.org/10.1007/BF00484597
33. Apter, M.: The Computer Simulation of Behaviour. Hutchinson, London (1970)
34. Thagard, P.: Rationality and Charity. University of Chicago Press, Chicago (1983)
35. Kosinski, J.: Being There. Bodley Head, London (1970)
36. Kmita, J.: Z metodologicznych problemów interpretacji humanistycznej. Państwowe Wydawnictwo Naukowe, Warszawa (1971)
37. Kmita, J., Nowak, L.: Studia nad teoretycznymi podstawami humanistyki. Państwowe Wydawnictwo Naukowe, Poznań (1968)
38. Dennett, D.C.: Content and Consciousness. Routledge and Kegan Paul, London (1969)
39. Drayson, Z.: The personal/subpersonal distinction. Philos Compass **9**(5), 338–346 (2014). https://doi.org/10.1111/phc3.121
40. Dennett, D.C.: Intentional systems theory. In: Walter, S., McLaughlin, B. (eds.) The Oxford Handbook of Philosophy of Mind, pp. 339–350. Oxford University Press, Oxford (2009)
41. Ryle, G.: The Concept of Mind. Hutchinson's University Library, London (1949)
42. Bennett, M., Dennett, D.C., Hacker, P., Searle, J.R.: Neuroscience and Philosophy: Brain, Mind, and Language. New York (2003)
43. Brożek, B.: Philosophy in neuroscience. In: Brożek, B., Mączka, J., Grygiel, W.P. (eds.) Philosophy in Science, pp. 163–188. Kraków (2011)
44. Damasio, A.: Descartes' Error: Emotion, Reason, and the Human Brain. G.P. Putnam's Sons, New York (1994)
45. MacMillan, M.: An Odd Kind of Fame: Stories of Phineas Gage. MIT Press, Cambridge, Mass (2002)

46. Block, N.: Psychologism and behaviorism. Philos. Rev. **90**(1), 5–43 (1981)
47. Lem, S.: Summa technologiae (Trans. Zylinska, J.). University of Minnesota Press, Minneapolis (1974/2013)

Part I
Functional Non-teleosemantic Concepts of Representation: From Internalism to Externalism

In Chap. 2, I pointed out that the concept of mental representation is characterized by intentionality. The concept of an intentional strategy does not answer the question of where intentionality comes from, a property which allows representational mental states of the cognitive system to be about something and to have some content. Thus, the problem of content determination of representational mental states remains open.

In subsequent chapters, I address the question of what gives rise to the ability to represent. The question of the determination of content and reference of mental representations requires defining how the cognitive system is capable of representing functions. In Chap. 3, I begin discussing naturalistic concepts that characterize the content of mental representations as conceptual roles played by representations in the cognitive system. I refer to these theories as "functional" because mental representations are considered functionally, or from a design stance as understood by Dennett. All such concepts solve problems that cannot be resolved by adopting an intentional stance alone. This is not to say that the concept of function plays a key role in these theories, whether understood one way or another; some theories do not refer to it at all, as evidenced by the causal theory of reference or Davidson's theory of meaning as conditions of truth. However, their common task is to identify and generally characterize non-intentional factors determining content.

I begin the characterization of functional concepts of mental representation with internalism (Chaps. 3 and 4). Internalism expresses the intuition that content properties are determined within the cognitive system. As such, they are independent of environmental factors. In Chap. 3, I analyze single-factor concepts of narrow content, and in Chap. 4, two-factor concepts. In both chapters, I show that internalism has certain advantages, emphasizing the significance of intentional properties in representational explanations. The drawback of internalism, however, is its failure to consider reference and conditions of satisfaction within the conception of content (in the single-factor version) or to deny their explanatory significance (in the two-factor version). Therefore, in Chap. 4, I reject the argument for narrow content as distinct from broad content.

In Chaps. 5 and 6, I analyze non-teleosemantic externalist conceptions: in Chap. 5, externalism in the informational-causal conception, and in Chap. 6, the truth-conditional conception. I point out that neglecting intensionality leads to conceptual difficulties. I summarize Part I by stating that non-teleosemantic conceptions of mental representations do not fulfill the desideratum of the naturalization of intentionality (II.2). This is because they neither specify the determinants of reference nor the contents (desideratum II.2.1), nor do they demonstrate why mental representations can bear content (desideratum II.2.2).

Chapter 3
Internalist Single-Factor Conceptions of Representation

In this chapter, I introduce the concept of narrow content (Sect. 3.1) to discuss the conception of narrow content through the example of Fodor's methodological solipsism (3.2). I point out that Fodor's formalist stance, which posits that content is reducible to the formal properties of mental representations, eliminates (at least—as I show in Sect. 3.3.4, in Stich's interpretation) semantic properties in favor of syntactic ones. Furthermore, this leads to conceptual difficulties, which I identify by drawing on the arguments of John Searle, Steven Harnad (Sect. 3.3), and Tyler Burge (Sect. 3.4). I illustrate these problems with selected examples from mind and brain sciences (Sect. 3.5). Consequently, I reject internalism in favor of externalism. In the remainder of the book, I analyze externalist conceptions.

3.1 Philosophical Concept of Intension and Extension

It is commonly assumed that there is an analogy between linguistic meaning and the content of mental representations [1]. Mental representations and linguistic expressions are traditionally attributed with intension or connotation[1] (meaning for linguistic expressions, content for mental representations) and extension (a mental representation can be about something, be true or false about some object, or be true or

[1] Connotation is the minimal set of properties of the designata of a linguistic expression, such that anyone who knows the language to which a given expression belongs can recognize the designatum of that expression by identifying the features of the object denoted by the expression. This understanding of connotation is traditionally used interchangeably with the concept of intension [2]. However, these concepts are sometimes distinguished. This happens with a slightly different understanding of intension. In the Carnapian tradition, intension is a function of linguistic expressions into consistent sets of sentences [3], while in the tradition of two-dimensional semantics, it is a function into possible worlds [4]. This function assigns a name to an extension (in a given possible world). For example, in our possible world, the terms "Evening Star" and "Morning Star" have the same intension across contexts, but different connotations.

K. Bielecka, *What is Misrepresentation?*, Studies in Applied Philosophy, Epistemology and Rational Ethics 71, https://doi.org/10.1007/978-3-031-84375-4_3

false simpliciter). The justification for this analogy is the correspondence of language and thought with reality; both linguistic expressions and mental representations refer to and describe reality.

An expression is extensional when its denotation is a function of the denotations of the components of that expression [5]. In extensional contexts, substituting any component of an extensional expression with another component with the same denotation does not change the logical value of the entire expression. Therefore, knowing the denotations of the members of an extensional expression is sufficient to determine the logical value of the entire expression. I will illustrate this with the example of two sentences:

(a) George W. Bush voted for Barack Obama during the 2012 elections in the USA.
(b) The 43rd President of the United States voted for Barack Obama during the 2012 elections in the USA.

Sentence (b) was created by replacing one of the members of the extensional expression, "George W. Bush," with a member of the same denotation, "the 43rd president of the USA," yet the logical value of the sentence remains unchanged. Both expressions refer to the same individual, the former president of the USA, George W. Bush.

An expression is considered 'intensional' or 'referentially opaque' if and only if it is not 'extensional.' Within intensional contexts, the substitution of one part of a sentence with another that holds the same logical value can change the logical value of the entire sentence. Intensional expressions include sentences of the form "x thinks that p," "x knows that p," "x desires that p," "x believes that p," and so on. Suppose George W. Bush voted for Barack Obama, but John does not know that George W. Bush is the 43rd president of the USA. I will illustrate this with an example:

a*) John is convinced that George W. Bush voted for Barack Obama during the 2012 elections in the USA.
b*) John is convinced that the 43rd of the United States voted for Barack Obama during the 2012 elections in the USA.

Both phrases "George W. Bush voted for Barack Obama during the 2012 elections in the USA" (in sentence a*) and "The 43rd President of the United States voted for Barack Obama during the 2012 elections in the USA" (in sentence b*) have the same logical value, yet sentence a* may have a different logical value than sentence b*, if John believes that the 43rd president of the USA and George W. Bush are two different people.

3.2 The Semantics of Conceptual Roles in the Naturalistic Version

The semantics of conceptual (functional, computational, inferential) roles is a heterogeneous array of distinct positions, which can be divided into naturalistic [6–11, 17] and normative [12–16] stances. Naturalists strive to elucidate the notion of the content of mental representations in non-semantic terms. Proponents of the normative position do not consider the explication of content in semantic terms to be a flaw (it is, in fact, a characteristic feature of such positions). However, their goal is different—to create a normative theory of meaning. This is precisely why I limit myself to the naturalistic semantics of conceptual roles.

The naturalistic conception of the content of mental representations, which elucidates content in terms of their functional roles in the cognitive system, is a promising attempt to create semantics for psychology. The functional role is typically conceived as the role a representation plays in the cognitive processes of the cognitive system, for instance, in perception, thought, or decision-making [17].

The definition cited from Block of the conceptual role is, however, quite general. It can be accepted by both internalists, who analyze the concept of narrow content, and externalists, who examine the concept of wide or broad content. Narrow content is limited to the functional role within the cognitive system, while wide content also takes into account the context and environmental factors in which the cognitive system is situated (I discuss the concepts of wide content in subsequent chapters).

3.2.1 In Search of Narrow Content. The Cognitive Character of Content

According to internalists, the concept of narrow content suffices to describe, explain, and predict the intentional states of all cognitive systems. They believe that the concept of wide content is flawed and unhelpful in psychology because it seems too dependent on the circumstances of content emergence and not enough on the structure of the cognitive system itself. To assess these theses, I will analyze the concept of narrow content, the reasons cited in its defense, and the intuitions underlying internalist views.

Internalists point to the mind's capacity to think about things that do not occur. I will illustrate these intuitions with the example of John standing at a bus stop waiting for a bus. John, having an incorrect timetable, can freely think about the bus, even though the vehicle does not and will not arrive. Moreover, the same thought about the bus can occur in the boy's head at different times and in different places: both at the bus stop and after returning home. Thus, it would seem that thoughts retain their content regardless of context (place, time). John can also think at any time about a mountain of gold, even though he knows perfectly well that such a mountain does not exist. Therefore, thoughts maintain their content even when they do not refer to anything real.

The concept of narrow content allows for the independence of thought content from factors such as reference and conditions of satisfaction. Thus, internalist Gabriel Segal defines it as cognitive content as distinct from referential/truth-conditional content [18].

Internalism stands in opposition to externalism, including the informational-causal Wisconsin-style theory of reference (see Chap. 5), which rejects the autonomy of content with respect to reference (more on the causal theory of reference in Chap. 5). The idea of extreme externalism was expressed by Hilary Putnam in his thought experiment involving Twin Earth [19]. Putnam demonstrates that intension does not determine extension. Imagine that on a distant Twin Earth, the liquid we call "water" does not refer to the chemical substance H_2O, but to XYZ. Water from Earth is indistinguishable from water on Twin Earth: it looks and tastes the same. Imagine that before the discovery of the chemical structure of water, before 1750, there lived on Earth a man named $Oscar_1$, who had a twin brother on Twin Earth, $Oscar_2$. The Oscars cannot distinguish XYZ from H_2O, and for both, the intension of the term "water" is the same: a colorless, drinkable liquid. Putnam argues that the meaning of the term "water" used by $Oscar_1$ is not identical to the meaning of the term "water" on Twin Earth, because the denotation of these terms is different. Thus, it is not the case that intension fully determines extension, as objective environmental states are also significant for meaning.

Internalists claim that there is a shared meaning of the term "water" on both planets, which is not captured by the externalist thesis, according to which there are different contents of the same concept on both Earths. This thesis, in the view of internalists, overlooks an essential aspect of mental content, which allows one to assert the same thing about the same object, regardless of external factors.

3.2.2 The Thesis of the Determination of Narrow Content as a Thesis on Supervenience

According to Block's definition, proponents of internalism offer a characterization of content in terms of only those causal roles that are inferential roles within an individual cognitive system. Narrow content is thus an element of the internal structure of an individual cognitive system, participating in its cognitive life, including its reasoning, decision-making, etc. [17]. The fact that it concerns solely the internal structure of an individual cognitive system is emphasized in Segal's definition (characterizing narrow content in terms of local supervenience), according to which narrow content is a property of the content of mental representations, completely independent of external factors relative to the system. Internalism assumes that only the internal structure of an individual cognitive system can determine narrow content.

The content of mental representations is entirely determined by the intrinsic properties of the subject or cognitive system. An intrinsic property is one that an object possesses (or does not possess) regardless of what occurs outside of that object [20]. An example of an intrinsic property is "a square having four equal sides": a square always has four equal sides, no matter what happens beyond it. However, the property of being a living organism is not an intrinsic property of that organism, because the organism would not have this property if, for instance, it did not breathe oxygen. Thus, the intrinsic properties of a cognitive system are those properties of the system's internal structure that remain independent of external factors. I will use the term "microstructural properties" to refer to such properties. Microstructural properties are the properties of the internal structure of the cognitive system and its parts (and the relations between them)—as opposed to macrostructural properties, which also include relations with the environment and other entities.[2]

The thesis on the determination of narrow content, expressed in terms of local supervenience, asserts that content properties necessarily depend asymmetrically only on the microstructural properties of the cognitive system. Segal characterizes narrow content in terms of local supervenience on the cognitive system's microstructure. He claims that microstructural properties suffice to determine the neuronal and computational properties of the system, that is, the narrow content:

> Fix an object's microstructure and you fix its atomic and molecular structure, its neurological and computational properties, and so on. [18: 14]

The characterization of the determination of narrow content in terms of local supervenience allows for a more precise explication of how content is determined in internalism, while also highlighting a problem with such a conception of determination. Local supervenience, in itself, does not provide a method for defining properties independent of the external environment of the system, yet it is on these properties that narrow content depends. On the contrary, it only excludes a certain group of concepts (externalist). Moreover, a consequence of the thesis on local supervenience is that all interactions of the cognitive system with its surroundings are irrelevant to content, which is fully reducible to the properties of the system's microstructure.

Here, ontological reduction comes into play. This is a relation between elements of the real world, such as objects, events, or properties. It occurs when relations such as elimination, identity, composition, realization, or supervenience take place [22]. Since internalism speaks of the supervenience between properties, this is an example of such ontological reduction.

[2] I have adopted the terms "microstructural" and "macrostructural" from Robert Poczobut, who formulated the thesis on supervenience in terms of micro- and macrostructure [21].

3.3 Fodor's Methodological Solipsism

I will now examine Fodor's defense of the concept of narrow content within his methodological solipsism [23, 24]. In Fodor's conception, narrow content arises from the reduction of semantic properties to syntactic properties and to the semantic properties of innate concepts. However, this reduction is not a full naturalization, as it is unclear how the content of innate concepts is determined. Thus, only a complete reduction of content to syntactic properties is fully consistent and naturalistic, but such an approach to narrow content deprives it of content properties. Consequently, a syntactic understanding of content is—in contrast to Fodor's view—inadequate for psychology and cognitive sciences. The semantics of conceptual roles, which would adopt such an understanding of narrow content, would therefore not be a semantics for psychology.

3.3.1 Introduction to Fodor's Solipsism. Motivations

Fodor's methodological solipsism is a key position for the development of the concept of narrow content in psychology and the philosophy of psychology. Within this framework, Fodor attempts to defend individualistic psychology as the proper way to practice psychology.[3] He responds to Putnam's objections against the classical thesis that intension determines extension.

In methodological solipsism, narrow content is characterized by inferential roles that are syntactic and computational.[4] Fodor argues that knowledge of the internal properties of a representation, especially its formal properties, is sufficient to describe its content. He claims that this would be enough to capture all content properties relevant to cognitive psychology.

Fodor further argues against the existence of a sufficiently good causal theory of content determination. He asserts that describing the world in causal terms would require complete knowledge of natural kinds. Without this, one could not speak of the actual reference of representations to the world.

In summary, Fodor argues for the adoption of narrow content because only narrow content serves to defend scientific psychology. The nonexistence of a sufficiently

---

[3] Later on, Fodor did not tie individualism in psychology and methodological solipsism together as strongly. Moreover, he distinguished between methodological solipsism and methodological individualism [25]. According to methodological solipsism, mental states are individuated without reference to semantic valuation, a (externally) relational taxonomy of mental states is methodologically inadmissible. Methodological individualism, on the contrary, allows for the relational individuation of mental states with the proviso that a property of a mental state is included in the taxonomy of mental states only if it has a causal impact [26].

[4] Identifying inferential roles with causal ones is associated with the classical computational vision of functionalism. Inferential roles characterized syntactically are those that participate in the computational architecture of the mind as a computer [27, 28].

good concept of reference determination leads Fodor to consider reference as methodologically irrelevant in psychology. Consequently, he also questions the usefulness of the concept of broad content in psychological explanations. He believes that the intuition that "mental processes have access only to formal (nonsemantic) properties of the mental representations" [23: 63] is not only advantageous but actually necessary, as "the contrary view is not only empirically fruitless but also conceptually unsound" [23: 68].

Fodor's radical proposal is evident in his argument against "naturalistic psychology," which he understands as individuating mental states in a transparent (extensional) manner, that is, broadly:

> Generally, then, a naturalistic psychology would attempt to specify environmental objects in a vocabulary such that environment/organism relations are law-instantiating when so described. But here's the depressing consequence again: we have no access to such a vocabulary prior to the elaboration (completion?) of the nonpsychological sciences. "What Granny likes with her herring" isn't, for example, a description under which salt is law-instantiating; nor, presumably, is "salt." [23: 70]

In other words, Fodor believes that not appealing to environmental properties in semantics is not a matter of convenience or methodological elegance, but of sheer necessity—psychology that refers to the environment is simply unscientific because it describes reality in a merely commonsensical way, omitting the laws of nature, and thus lacks explanatory power.

Fodor's argument is valid, but unsound: if indeed a good concept of reference determination were impossible, then psychology would have to be content with narrow content. The conclusion indeed follows, but from a false premise. Psychology, in his view, is impossible because only intrinsic properties have causal powers [25]. This assumption of Fodor's is false (numerous counterexamples are provided by Wilson [29]. Moreover, one can doubt whether living biological organisms have intrinsic properties at all, since they cannot exist independently of the state of the environment, and yet it is difficult to deny biological organisms causal powers.

Fodor assumes that only natural kinds can be causes [23]. Moreover, without complete knowledge of natural kinds, we cannot know causal relations. In this way, he further justifies the alleged impossibility of a naturalistic conception of reference. However, the claim that without full knowledge of natural kinds one cannot know causal relations is false, as it contradicts common practice: there is nothing absurd in attributing a cause without knowing all the properties of natural kinds and the laws of nature.

3.3.2 Narrow Content in Methodological Solipsism

According to Fodor, the relation of representation has two interconnected components. It consists of a relation with the propositional object described formally (syntactically) and a relation with the same object characterized semantically in terms of truth and reference. However, as Fodor adds, the formal features of narrow content are sufficient to semantically distinguish different representations. Fodor is thus an advocate of the formalist stance in the theory of content, according to which the formal properties of signs (e.g., shapes, structure) and syntactic rules are sufficient for characterizing content.

Fodor conceives of the mind as a so-called machine with an oracle, where by "oracle," following Turing, he means a decision procedure that answers some question in a non-computational (non-algorithmic) way. According to Fodor, the role of the oracle is played by states of the environment perceived perceptually:

> The point is that, so long as we are thinking of mental processes as purely computational, the bearing of environmental information upon such processes is exhausted by the formal character of whatever the oracles write on the tape [of the machine - note K.B.]. In particular, it doesn't matter to such processes whether what the oracles write is *true*; whether, for example, they really are transducers faithfully mirroring the state of the environment, or merely the output end of a typewriter manipulated by a Cartesian demon bent on deceiving the machine. [23: 65]

Fodor claims that a formal difference determines a functional difference, and this, in turn, determines a causal difference:

> The form of explanation goes: it's because different content implies formally distinct internal representations (via the formality condition) and formally distinct internal representations can be functionally different; can differ in their causal role. Whereas, to put it mildly, it's hard to see how internal representations could differ in causal role unless they differed in form. [23: 68]

For Fodor, this is a pragmatic argument for the replacement in explanations of semantic properties with formal ones.

Fodor, much like Davidson, champions folk psychology, hence he emphasizes that sentences about beliefs, which are opaque contexts, are pivotal in folk psychology. He even asserts that folk psychology requires nothing beyond explanations of representations in referentially opaque contexts, where generalizations concern how people understand objects that fall under judgments to which they hold propositional attitudes.

Sentences occurring in opaque contexts differ not only in content but also in form. This is precisely why such sentences do not conflict with the first assumption that only internal structural properties have causal powers. According to Fodor, formal properties suffice for an adequate description of content, such as beliefs, even in opaque contexts. I intend to argue against Fodor's concept of narrow content, according to which a formal difference is enough to explain a difference in intension.

3.3.3 Innateness of Concepts

Fodor has his own concept theory, which is a philosophical interpretation of the classical theory of concepts, enriched by the hypothesis of innateness. In the classical theory, a concept is a class representation, encompassing all the relevant properties of that class. The criterion of relevance is repeatability—an essential feature is one that characterizes all objects belonging to that class, or in other words, it is a common feature of those objects belonging to the class [30]. Maruszewski's definition does not differ much from the classical definition of connotation, according to which connotation is the specificity of objects, a property that belongs to all and only z's (see also [5: 107].

The innateness hypothesis posits that our basic conceptual apparatus is innate.[5] We possess a language of thought, or an internal code, composed of basic concepts to which other concepts are reducible, and we are also capable of performing the appropriate combinatorial operations [28]. For example, the non-basic concept BORING BOOK[6] is composed of the concepts BOOK and BORING. An important argument Fodor presents for the language of thought in psychology is the productivity of thought. A suitably productive system is language. Fodor must still justify why language, rather than another productive representation system, is innate. This is where the key argument for the innateness of language comes into play.

Fodor [28], in his analysis of psychological theories of concept learning, observes that all such theories treat the learning of concepts as a process of formulating hypotheses. Hypotheses can only be posited in a language, therefore, for a child to formulate them, it must possess an innate language of thought, because before learning concepts, it must be able to formulate these hypotheses somehow. Hence, it is language, rather than any other productive representational system, that is innate: one cannot formulate hypotheses in any other system.

Let us return to the issue of productivity. It is evident that we can think a potentially infinite number of different thoughts; similarly, we can utter a potentially infinite number of sentences, such as "Giraffes do not play poker." According to Fodor, it is impossible to explain the productivity of language and thought without the assumption of the compositionality of language. It is precisely language that has a structure rich enough to be productive. Therefore, Fodor believes, thinking occurs in a linguistic medium. Fodor's concept thus fulfills desideratum II.3.1 regarding the inclusion of the productivity of the representational system.

A particular case of concepts in Fodor's theory are the aforementioned basic concepts, which cannot be further simplified. These include concepts describing simple sensory qualities, such as shapes and colors, as well as the simplest concepts

---

[5] The concept of innateness is ambiguous and can be understood in many different ways, especially since in the context of contemporary psychology one does not abstract from the biological substrate of cognition. In biology, it is not only unclear what is innate, but also what innateness consists of; philosophers of biology question the utility of this concept [31–33]. One explication of the concept of innateness in biology reduces it to a disposition for behaviors under normal conditions. However, this explication must go beyond narrow content, as normal conditions always manifest in the environment (more on the complications associated with the concept of normal conditions in naturalistic causal theories of content can be found in Chap. 5).

[6] In the literature on mental representations, it has been conventional to denote the names of concepts, as opposed to the names of terms in natural language, with capital letters (e.g., BOOK is a concept in thought, while "book" is the name of a word in the English language).

from colloquial language. Whether a concept belongs to the basic category is also determined by its structure: it must be a concept without which other concepts cannot be formed based on the compositional principle. For example, the concept BORING BOOK is a non-basic concept composed of the basic concept BOOK and the basic concept BORING, and the concepts BOOK and BORING cannot be further simplified and are concepts without which it would be difficult to form the concept BORING BOOK.

Methodological solipsism, by assumption, prohibits referring to the properties of the environment in explaining the determination of content of these concepts. Thus, they remain unexplained, and in essence, his theory can never elucidate how their content is determined. The content of basic concepts, therefore, remains an aporetic aspect of methodological solipsism; it is no wonder, then, that Fodor later rejects solipsism and adopts informational semantics [25, 34], and subsequently semantics based on asymmetric dependency [35]. Within the framework of methodological solipsism, the determination of content of basic concepts is logically impossible: if they are innate, they depend not only on the individual cognitive subject but also on biological inheritance. In an externalist approach, the determination of content of innate concepts remains a challenging problem, yet its resolution is not logically impossible.

The Hypothesis of an innate conceptual apparatus could remain consistent with the thesis on the local supervenience of narrow content on the formal properties of the conceptual apparatus, provided that the formal properties of this apparatus were the basis for the content of concepts. However, Fodor opposes the complete reduction of semantics to formal properties. One way out of the problem of determining the content of basic concepts, which allows for the preservation of methodological solipsism, is to treat it as a formalist conception, that is, purely syntactic. This would strengthen and radicalize Fodor's position.

Fodor does not endorse the strong thesis that all semantic properties reduce to syntactic ones, but he claims something weaker: namely, that formal properties are the most satisfactory indicator of content. I do not intend to argue with this: indeed, form is the simplest indicator of difference in content and, moreover, it performs excellently in many situations.

3.3.4 Methodological Solipsism and the Formalist Position

In this section, I analyze an even more radical position of Stich, who argues for a complete reduction of semantics to form and, consequently, for the elimination of the concept of representation from folk psychology. Stich points out exactly what the reduction of content to formal properties leads to. Stich's argument can be seen as a defense of the common concept of representation, characterized by both form and content that is irreducible to this form. Essentially, Stich shows that the internalist concept of content is not a concept of content at all. This means that internalism cannot naturalize intentionality.

Stephen Stich starts from a formalist interpretation of methodological solipsism, criticizing Fodor's argument for the necessity of adopting this stance. Stich's counterargument can be understood as directed against methodological solipsism as well as against psychology that utilizes the concept of representation. Stich's objection to Fodor's concept of narrow content is to demonstrate that it is divergent from the common concept of content and, consequently, cannot be used in defense of folk psychology. Thus, Stich presents arguments for rejecting any concept of content. Moreover, he opposes representationalism, as he believes—like Fodor—that computational psychology does not adhere to the principle of charity, and therefore does not take into account semantic properties of representations such as truthfulness.

Stich accuses Fodor of leading to undesirable consequences with the concept of narrow content as constructed by Fodor:

> First, most computational (or formal) mental state types will have tokens (either actual or possible) whose contents are radically different from one another, as judged by our "aboriginal, uncorrupted, pretheoretic intuition." Second, there will be some computational mental state types whose tokens can be assigned no content at all by our aboriginal intuition, though these "contentless" computational states will serve the purposes of the computational theory of mind fully as well as their contentful cousins. [24: 97]

Firstly, Stich notes that the concept of narrow content leads to the ascription of the same content in intuitively different cases. Secondly, he emphasizes that the approach to narrow content, typical of methodological solipsism, leads to another undesirable consequence: it does not prohibit the ascription of content to beliefs very distant from our own, even if they violate common intuitions regarding their content.

Cases of the first kind are analogous to Putnam's examples from Twin Earth. Let me recall one of them. Stich introduces Fodor from Yon (an analogue of Putnam's Twin Earth), who states, similarly to Fodor from our Earth: "Jimmy Carter is from Georgia." On Twin Earth, even further removed from ours than Yon, there could exist a twin Fodor, uttering the same sentence "Jimmy Carter is from Georgia," and yet, according to Fodor's conception of narrow content, this sentence on Twin Earth should be ascribed the same content. Stich observes that this is completely inconsistent with the common intuition, which suggests that different Fodors, depending on their location, are talking about different Jimmy Carters, depending on the places they originate from (assuming that on Yon and Twin Earth there are also places named "Georgia"). I agree with Stich that ascribing content to a belief should take context into account. Undoubtedly, defending the reference of individual proper names requires agreement with such a version of the causal theory of content that is inconsistent with methodological solipsism. The dependence of content on context is one of the reasons arguing for the adoption of wide content.

The concept of narrow content should also be attributed to things or beings—and this falls under the second type of case—to which the principle of charity would forbid us to attribute it. Stich illustrates this with the example of robots whose beliefs are so distant from ours that one could never agree that—according to the principle of charity—their representational states have any content. If there were a robot simulating a human, possessing beliefs that are largely inconsistent with one another, the principle of charity would not allow us to attribute content to it, since in the case of such a robot one could not speak of neither the truth nor the accuracy of its "beliefs." However, Fodor lacks a method to deny the sensibility of these "beliefs."

In summary: Stich shows that the formalist stance does not lead to the defense of folk psychology but rather to its rejection. Although he agrees with Fodor that the formalist position suffices to describe content, he goes a step further, claiming that the concept of representation should be altogether discarded from folk psychology. At the same time, the concept of narrow content should be rejected, leaving only the concept of form. I agree with Stich's argument against the concept of narrow content as content reduced to formal properties. This reduction does not allow for an accurate description of contexts in which expressions differ in content but not in form. However, I propose to treat Stich's argument—contrary to his intentions—as a warning against too hasty a reduction of content to form, as well as against the elimination of the concept of mental representation.

Narrow content in the context of methodological solipsism fails to differentiate between the different meanings of ambiguous representations when their form is the same. This group includes homonyms (such as 'bank') and representations whose meaning depends on the environment (Putnam's Twin Earth example). Their form is the same, but the content differs because: (a) the intension and extension are different (in the case of homonyms) or (b) the extension is different (for those representations whose content depends on the environment). The formalist position could handle homonyms at the level of expressions by effectively denying their existence: by isolating sentences in which the word 'bank' means 'slope' from those where it means 'financial institution,' due to the inferential roles these words play in sentences, which determine the meaning of 'bank.' The formalist position has a much harder time dealing with homonyms at the level of entire sentences, where the pragmatic context plays a decisive role—the sentence "You have big feet!" stated in a shoe store asserts the fact that someone has feet of a large size, while the same sentence stated in the presence of a person with small feet becomes ironic; one must also remember the role played by intonation—a sentence spoken in a contemptuous tone can be offensive [36]. In the case of homonyms at the level of entire sentences, resorting to inferential roles is practically unfeasible—it is impossible to distinguish sentences so strongly context-dependent solely at the syntactic level, thus referring only to formal properties.

However, the formalist position can account for the semantic difference between the sentences "George W. Bush voted for Barack Obama during the 2012 U.S. elections" and "The 43rd President of the United States voted for Barack Obama during the 2012 U.S. elections," since the term "George W. Bush" has a different form than "The 43rd President of the United States."

The formalist position is effective in many interesting cases where the form of the carrier corresponds to the content. It would be a mistake to completely abandon considering formal properties in the conception of content. However, for a formal conception to fully replace the semantic one, it should explain the troublesome cases described above.

An adequate concept of content should not be merely a formal concept of content, as the formalist position is impotent in contexts where reference clearly determines content. Oscar's statement on Twin Earth about water is considered different from Earthly Oscar's statement, precisely because "water" refers to a different chemical substance on Earth than it would on Twin Earth. It is therefore reasonable to adopt the assumption that an adequate concept of content should not be merely formal, as reducing content to formal properties leads to the failure to meet the desideratum of determining reference and content (II.2.1). This assumption seems necessary to naturalize intentionality, which is expressed by desideratum II.2.

3.4 Arguments Against the Formalist Position

3.4.1 The Chinese Room Thought Experiment

The most famous critique of proponents of a formal approach to representations, in this case symbolic representations, is found in John R. Searle's Chinese Room thought experiment [37]. Searle, who does not speak Chinese, is locked in a room with a stack of Chinese texts ("script") along with additional sets: a set of rules (acting as a program) correlating the second file with the first (called by Searle "story") and a set of instructions in English (questions), allowing the correlation of elements of the third file with the first two. These instructions indicate how to refer certain Chinese symbols of given shapes, responding to those shapes presented in the third file. Both the program and the instructions for responding to questions are for Searle purely syntactic transformations of symbols. Searle, in the room, is to answer questions asked in both English and Chinese. As it turns out, his responses to the Chinese questions are indistinguishable from those of a native Chinese speaker, and furthermore, the responses to the Chinese questions are just as good when viewed from the outside as the responses to the English questions, even though in the first case Searle's responses are merely uninterpreted symbols.

Searle's thought experiment touches on many philosophical issues, including consciousness, problems concerning the computational capacities of the brain, and mental representations. Given the focus of this work, I will concentrate on the latter, and within that scope, on the issue currently of interest to me: the possibility of characterizing content solely in formal terms. To this end, I will refer to one of the many formulations of Searle's argument, and within it, to the relationship between form and content [38]. The argument takes the following form:

(a) Programs are purely formal.
(b) Minds (at least human ones) possess semantics, mental content.
(c) Syntax alone does not constitute content nor is it a sufficient condition for content.

Programs themselves neither constitute nor are a sufficient condition for the operation of the mind [39].

Many philosophers question the validity of Searle's argumentation [40, 41]. However, I believe that the essence of Searle's argument, the thesis that syntax is neither identical with content nor sufficient to describe semantic content, points to a problem that remains real for formalistic concepts of mental representation. This is the case regardless of Searle's concept of intentionality [42], which is based on intuitions about consciousness. In Searle's concept, intentionality is inseparable from consciousness, and therefore one cannot speak of intentionality at the subpersonal level.

3.4.2 The Systems Reply

The systems reply to Searle's experiment amounts to saying that although Searle, as a person in the room, does not understand Chinese, the entire system does [37]. Proponents of the systems reply argue that the fact that the person in the room does not understand Chinese does not mean that the system does not understand it. Meanwhile, Searle has not demonstrated anywhere that the entire system does not understand. His opponents claim that Searle makes a mistake by conflating a part of the system with the whole: just because some part of the system does not understand, it does not mean that the entire system does not understand. Searle would only be correct if understanding could be divided into parts like an apple: mass pertains to the apple, and thus to its individual parts. Searle must indeed assume that every part of the system must think. If, therefore, a human being understands, then so does, for example, the stomach or liver,if the stomach did not understand, then the human would not understand either [43].

In response to this criticism, Searle argues that according to his opponents' arguments, a system that has memorized rules incomprehensible to itself, along with a piece of paper, constitutes a thinking system, which would be absurd. Searle asserts that there must exist a difference between 'authentically mental' systems and those that are not, and that the system itself must be able to determine this difference. Such a system, he believes, is characterized by biologically 'built-in' intentionality. The problem posed by Searle is profound, but the solution is unsatisfactory. This is because Searle a priori cuts off an interesting demarcation problem between thinking systems and other systems, without showing what their distinctness actually consists of. The assertion that systems are distinct is decidedly insufficient.

The Chinese Room recalls the argument against mechanism known from Leibniz's thought experiment with the mill. Leibniz writes about the nature of the mind:

> Moreover, it must be confessed that *perception* and that which depends upon it are *inexplicable* on *mechanical grounds* that is to say, by means of figures and motions. And supposing there were a machine, so constructed as to think, feel, and have perception, it might be conceived as increased in size, while keeping the same proportions, so that one might go into it as into a mill. That being so, we should, on examining its interior, find only parts which work one upon another, and never anything by which to explain a perception. Thus

> it is in a simple substance, and not in a compound or in a machine, that perception must be sought for. [44: 227–228]

Searle argues similarly to Leibniz that without consciousness, understanding is impossible (Leibniz claimed that there is no thinking without perceptions). However, Searle himself, not being a dualist, recognizes that a biological brain suffices for understanding (intentionality). Meanwhile, one can imagine a brain as large as a mill—and still, no consciousness would be visible if one were to enter this mill. Thus, the intuition that mechanism is insufficient for intentionality or perception may be mistaken. It is worth noting that the Chinese Room itself could be reduced to the size of a ball and realized through the human brain [40, 45]. Worse still, Searle never justified his thesis that intentionality is impossible without consciousness. The only support for this thesis is a thought experiment in the style of Leibniz, which, however, generates intuitions contrary to Searle's own position if applied to the brain. I reject Searle's thesis on the role of consciousness, as it remains unjustified and leads to conceptual tensions.

3.4.3 A New Version of the Thought Experiment with the Chinese Room

In light of the problem of content in concepts of mental representation that interests me, I will briefly present Searle's experiment in a slightly modified version by Stevan Harnad; here, I also separate the problem of intentionality from the problem of consciousness. Harnad better outlines the issue of the relationship between the formal symbol and the referent of mental representations, which in artificial intelligence is called the "symbol grounding problem" [46].

Harnad's experiment has two versions. In the first, Harnad describes learning Chinese as a first language solely with the aid of a Chinese-to-Chinese dictionary, while in the second, he describes learning Chinese as a second language, which he considers feasible, though difficult. I will refer to the first version.

In Harnad's variant, the Chinese Room thought experiment concerns the grounding of symbols in something other than meaningless other symbols. According to Harnad, it is not other symbols that constitute meaning, but rather relations with the world. The problem, therefore, is not whether it is possible to translate a specific language of a given linguistic form into another language of a different linguistic form, but how to relate the linguistic form to the world.

Searle's argument, as seen more clearly in Harnad, boils down to the fact that syntactic properties are insufficient to capture semantics, as reference and truth value are also needed for a proper description of semantics. Thus, the argument's edge is directed against the formalistic version of internalism. The formalistic stance fails to capture reference. As Fodor demonstrates with sentences in intentional contexts, based on the formal properties of the carrier alone, much can be inferred about the truth or falsity of representations (though not everything, as shown by Tarski and

Gödel [47]). However, the formalistic stance does not explain why a representation can be a carrier of truth. Even if one accepts the formalistic assumption that mental representations have a syntax that is causally significant, it is not at all clear that these representations can be true or false. Moreover, under formalistic assumptions, it is not at all clear what physical structure could be considered a syntactic structure and why certain syntactic constructions should correspond to falsehoods and others to truths. Thus, the formalistic position does not meet the desideratum of a content carrier (II.2.2).

3.5 Internalism and the Determination of Intentions Through Learning

Internalists have a problem explaining action, which is related to a psychologically implausible conception of concept learning. Fodor assumed that there are necessary and sufficient conditions for possessing concepts and that the concept of narrow content should serve a psychologically credible theory of learning [23]. However, such an approach to learning does not allow for the determination of intentions as a result of learning.

Burge's argument [48] targets the concept of narrow content and is predicated on the possibility of concept acquisition. In this example, we encounter a thought experiment constructed similarly to Putnam's experiment, in which Burge, however, argues for the role of social context in the determination of content, rather than determination by physical facts. In the experiment, we compare an individual whose physical states from birth to the present are identical, but occur in two different situations distinguished only by the linguistic community or social environment. Consequently, in these situations, the individual uses the concept of arthritis differently: while in the actual situation the person has a good understanding of the concept's extension, in the counterfactual scenario, the term "arthritis" is used to denote a disease that can affect both muscles and joints. According to Burge, the extension of the concept depends on the social context in which a person is raised and in which the concept is used. A proponent of narrow content might respond that in the second situation, the other person has only a vague understanding of arthritis, knowing only that it is a certain disease, but is mistaken about what kind of disease it is. However, the consequences that proponents of narrow content draw are too far-reaching. They claim, namely, that in the counterfactual case, one cannot attribute to the person knowledge of the concept's extension, since it is not known what it describes.

Usually, concept acquisition occurs over time, and conceptual understanding is graded. However, in the classical conception, one possesses the same concept only when it is fully mastered; therefore, a concept of arthritis that is not fully grasped is not yet the concept of arthritis. If this is the case, then it must be acknowledged that if we do not know the necessary and sufficient conditions for the application of concepts (and Fodor himself argues that we do not know them; cf. [49]), then we do not know the concepts themselves. Without the assumption of innateness, this leads to a very peculiar consequence. It is not easy to identify the necessary and sufficient conditions for the application of concepts such as "game," "chair," or "animal," and since they were not innate, then on the grounds of the classical theory of concepts, it would have to be said that we do not know them at all. However, if they are innate, then their content is not determined individually. Here again, we encounter a fundamental aporia in Fodor's conception: his nativism excludes methodological solipsism, as innate concepts must have content determined by factors that do not supervene locally, that is, they have content that extends beyond narrow content.

What an internalist such as Fodor accuses Burge of exposes the weaknesses of Fodor's internalism. An internalist could, of course, abandon both the classical theory of concepts and nativism, but then Burge's argumentation would strike him. He would have to agree that the key factors determining content are social.

3.6 Application of Fodor's Concept to Examples from Mind and Brain Sciences

I will now apply Fodor's concept of narrow content to psychological practice and examine some of the consequences it has for explanations in the sciences of the mind and brain. I omit mental imagery as an example of structural representations, since during his methodological solipsism phase, Fodor adopted Pylyshyn's views on the status of mental imagery. In his internalist period, he believed there was no reason to postulate representations other than linguistic ones, which mental images are. However, when he accepted structural representations, he became an externalist [50].

Rombix

The experiment with the concept ROMBIX shows that rombix can be recognized based on features of shape (a diamond on a square background) and color (the diamond is red), thus based on the connotations of the concept. Lewicki's concept is meant to explain how this happens, using the process of abstraction (negative and positive).

In Fodor's theory, the concept ROMBIX is a non-basic concept, composed of three basic concepts: RED, DIAMOND, and SQUARE. The theory of learning the complex concept ROMBIX would look almost the same as in Lewicki's approach: in response to stimuli, the subject receives a negative or positive response and then learns the given concept through the process of hypothesizing and confirmation (or refutation). The principle of hypothesizing and confirmation operates just like the principle of negative and positive abstraction in Lewicki's theory. No wonder, since Lewicki explains the processes of abstraction precisely in terms of hypothesizing and confirmation. Fodor assumes that there are necessary and sufficient conditions for possessing concepts, which are provided by the innate language of thought. As he claims, learning concepts is impossible without an innate language of thought. Through stimuli (in the example, these are cards shown by the experimenter) and punishments or rewards (which the experimenter symbolically gives to the subject), only the already innate basic concepts are activated, and these, thanks to the principle of compositionality, create a complex concept.

Fodor would likely respond to the question about the source of hypotheses or the set of hypotheses concerning concepts such as ROMBIX by referring to his own learning theory, which is a theory of content determination for non-basic concepts. Within this framework, the content of the non-basic concept ROMBIX arises from basic concepts RED, DIAMOND, SQUARE. These must be innate; however, it cannot be explained by what means they possess content.

Fodor's conception, however, allows for the description of the process of learning concepts through hypothesis formation, although nativism combined with internalism leads to an irresolvable aporia within naturalistic semantics: it is unclear how the language of thought could have semantic properties, and the assumption of internalism does not allow for the recognition that the language of thought acquires meaning in the interactions of the cognitive system with the environment or other cognitive systems. Nevertheless, Lewicki's research could be conducted under the assumptions of methodological solipsism. Thus, internalism emerges unscathed from this confrontation (leaving aside the issues of innateness).

Cognitive Map

I will now examine how Fodor's concept can be applied to a rat using a cognitive map. A cognitive map is not of a linguistic nature, but its structure is rich enough to ensure compositionality [51]. It is worth noting that initially, Fodor believed that all representations are symbolic in nature, and therefore he rejected the possibility of structural representations (such as representations understood in non-linguistic categories). In later works, Fodor acknowledges the possibility of geometric representations (see, e.g., [35]), and he uses the term "language of thought" more liberally to describe any computational medium of representation. Since this liberalization has no connection with the assumption of internalism or externalism, it might seem that at least in principle, Fodor has the opportunity to describe the content of cognitive maps in accordance with solipsistic methodology.

However, even the liberalization of assumptions about the language of thought is not sufficient for internalism to accurately describe research practice. To demonstrate this, I propose a thought experiment involving transferring a rat to a different maze.[7] This is a modified and adapted animal version of Putnam's thought experiment, based on the practice of comparative psychologists and ethologists who study rat behavior under various conditions.

Let's imagine the well-known Putnam's Twin Earth experiment concerning not an ordinary Oscar, but Oscar the rat. A researcher has confined Oscar in a maze where a nutrient solution is present, allowing the animal to survive. Oscar learns

[7] It is worth noting that experiments involving transferring rats to a different maze do take place in experimental practice and serve similar purposes as in my example [52].

the structure of the maze. Then, the researcher transports him to another, similar maze (analogous in this example to Twin Earth). The trained Oscar searches for the tasty nutrient solution, but in vain, as there is no such solution in the second maze. The simplest explanation for the rat's behavior would be that the rat represents the new maze as the old one and therefore searches for the watering place in a manner analogous to the old maze. However, it turns out that his representation is incorrect, and the nutrient solution simply isn't in the remembered location—this element of representation has an empty reference. Fodor, however, cannot use the concept of reference to explain the functioning of the representational mechanism in the rat or say that the rat refers to a different environment using a cognitive map. This would be inconsistent with methodological solipsism.

This thought experiment demonstrates that the concept of reference is necessary in the sciences of the mind and brain, and content alone is insufficient for conducting research on representational structures. Fodor's concept of narrow content dispenses with the concept of reference, thereby conflicting with scientific research practice. The concept of reference is essential for understanding representational-based adaptive behaviors, which are common among higher animals.

3.7 Summary

In this section, I have demonstrated the flawed reasoning of monofactorial internalist positions regarding the sufficiency of narrow content in the theory of representation. In summary: the characterization of narrow content either leads to ambiguity or strips the concept of content of its semantic properties. If by "narrow content" we mean—following Segal—a content property of mental representations that is completely independent of external factors relative to the system, then the concept of content remains elusive, and it is unclear what it could mean. While conceiving of content as partially independent of contextual factors allows for the preservation of content properties, it seems that conceiving of content in complete abstraction from external factors does not allow for the preservation of such properties. Meanwhile, Fodor's formalist stance, especially in Stich's radical interpretation, eliminates content properties in favor of syntactic properties.

Arguments presented by proponents of narrow content, such as the ability to formulate thoughts independently of context or thinking about non-existent objects, advocate for the concept of narrow content. Nevertheless, the concept of narrow content fails to account for reference (thus not fully meeting the desideratum II.2.1 on the determination of reference and content), nor conditions of satisfaction, without which it is impossible to understand how mental representations can be bearers of content (as expressed by desideratum II.2.2 on the bearer of content). By abstracting from reference and conditions of satisfaction, internalism does not fully satisfy desideratum II.3 on the proper concept of content, even though it considers the productivity of some representations (desideratum II.3.1). The lack of connection to reference and conditions of satisfaction means that the concept of narrow content does not allow us to speak of the adequacy of representations with respect to what they represent. Some of these representations concern the environment, which in a correct concept of content would allow explaining common adaptive behaviors in some animals, such as navigating the environment (e.g., through cognitive maps), as based on adequate representations of the environment. One-factor theories do not allow us to say, for example, that a predator made a mistake while hunting, searching for prey. In other words: these theories do not meet any of the two desiderata on fallibility (II.3.2, II.3.3).

The concept of narrow content does not support a promising theory of cognitive maps, and therefore does not allow for explanations of many cognitive processes, especially involving episodic memory. The example of cognitive maps is, however, a key empirical argument in favor of representationalism [53].[8]

References

1. Pitt, D.: Mental Representation. In: Zalta, E.N. (ed.) The Stanford Encyclopedia of Philosophy. Retrieved from http://plato.stanford.edu/archives/fall2013/entries/mental-representation/ (2013)
2. Copi, I., Cohen, C.: Introduction to Logic, 11th edn. Prentice Hall, Upper Saddle River, N.J. (2002)
3. Carnap, R.: Meaning and Necessity. University of Chicago Press, Chicago (1947)
4. Chalmers, D.J.: Epistemic two-dimensional semantics. Philos. Stud. **118**(1/2), 153–226 (2004). https://doi.org/10.1023/B:PHIL.0000019546.17135.e0
5. Jadacki, J.: Spór o granice języka: elementy semiotyki logicznej i metodologii. Wydawnictwo Naukowe, Warszawa (2002)
6. Field, H.: Logic, meaning, and conceptual role. J. Philos. **74**(7), 379–409 (1977)
7. Harman, G.: Reasoning, Meaning and Mind. Clarendon Press, Oxford (1999)
8. Horwich, P.: Meaning. Oxford University Press, Oxford-New York (1998)
9. Horwich, P.: Reflections on Meaning. Oxford University Press, Oxford-New York (2005)
10. Loar, B.: Mind and Meaning. Cambridge University Press, Cambridge-New York (1981)
11. Peacocke, C.: A Study of Concepts. MIT Press, Cambridge (1992)
12. Brandom, R.: Making it Explicit: Reasoning, Representing, and Discursive Commitment. Cambridge (1994)
13. Savigny, E.: The Social Foundations of Meaning. Springer, Berlin, New York (1988)
14. Sellars, W.: Empiricism and the philosophy of mind. Bobbs-Merrill, Indianapolis (1956)
15. Skorupski, J.: Meaning, use, verification. In: Bob, H., Wright, C. (eds.) A Companion to the Philosophy of Language, pp. 139–168. Blackwell, Oxford (1997)
16. Travis, C.: Unshadowed Thought: Representation in Thought and Language. Harvard University Press, Cambridge (2000)
17. Block, N.: Conceptual role semantics. In: Craig, E. (ed.) Routledge Encyclopedia of Philosophy, pp. 242–256. London, New York (1998)
18. Segal, G.: A Slim Book about Narrow Content. MIT Press, Cambridge (2000)
19. Putnam, H.: The meaning of meaning. In: Philosophical Papers: Mind, Language, and Reality, vol. II, pp. 215–271. Cambridge University Press, Cambridge (1975)
20. Yablo, S.: Intrinsicness. Philos. Top. **26**(1/2), 479–505 (1999)

[8] It is worth mentioning that rats are capable, to some extent, of predicting locations they will find themselves in, that is, thinking about states of affairs that have not yet occurred [54].

21. Poczobut, R.: Eksternalizm treści umysłowej a superweniencja. Kognitywistyka i Media w Edukacji **9**(1), 82–106 (2007)
22. van Gulick, R.: Reduction, emergence and other recent options on the mind-body problem. A philosophic overview. J. Conscious. Stud. **8**(9–10), 1–34 (2001)
23. Fodor, J.A.: Methodological solipsism considered as a research strategy in cognitive psychology. Behav. Brain Sci.. Brain Sci. **3**(1), 63–109 (1980). https://doi.org/10.1017/S0140525X00001771
24. Stich, S.P.: Paying the price for methodological solipsism. Behav. Brain Sci.. Brain Sci. **3**(1), 97 (1980). https://doi.org/10.1017/S0140525X00002016
25. Fodor, J.A.: Psychosemantics: the Problem of Meaning in the Philosophy of Mind. MIT Press, Cambridge (1987)
26. Heath, J.: Methodological individualism. In: Zalta, E.N. (ed.) The Stanford Encyclopedia of Philosophy. Retrieved from https://plato.stanford.edu/entries/methodological-individualism/ (2015)
27. Field, H.: Mental representation. Erkenntnis **13**, 9–61 (1978)
28. Fodor, J.A.: The Language of Thought. Thomas Y. Crowell Company, New York (1975)
29. Wilson, R.A.: Boundaries of the Mind. The Individual in the Fragile Sciences. Cognition. Cambridge University Press, Cambridge (2004)
30. Maruszewski, T.: Analiza procesów poznawczych jednostki w świetle idealizacyjnej teorii nauki. Wydawnictwo Naukowe, Poznań (1983)
31. Samuels, R.: Nativism in cognitive science. Mind Lang. **17**(3), 233–265 (2002). https://doi.org/10.1111/1468-0017.00197
32. Samuels, R.: Innateness in cognitive science. Trends Cogn. Sci.Cogn. Sci. **8**(3), 136–141 (2004). https://doi.org/10.1016/j.tics.2004.01.010
33. Griffiths, P.E.: What is innateness? Monist **85**(1), 70–85 (2002)
34. Fodor, J.A.: Semantics, Wisconsin style. Synthese **59**(3), 231–250 (1984). https://doi.org/10.1007/BF00869335
35. Fodor, J.A.: LOT 2: The Language of Thought Revisited. Clarendon Press, Oxford, New York (2008)
36. Dennett, D.C.: Consciousness Explained. Little, Brown and Company, New York (1991)
37. Searle, J.R.: Minds, brains, and programs. Behav. Brain Sci.. Brain Sci. **3**(3), 1–19 (1980). https://doi.org/10.1017/S0140525X00005756
38. Hauser, L.: Searle's Chinese box: debunking the Chinese room argument. Mind. Mach. **7**(2), 199–226 (1997)
39. Bishop, M., Preston, J.: Views into the Chinese Room: New Essays on Searle and Artificial Intelligence. Oxford, New York (2002)
40. Chalmers, D.J.: The Conscious Mind: in Search of a Fundamental Theory. Oxford University Press, New York (1996)
41. Dennett, D.C.: The Intentional Stance. MIT Press, Cambridge (1987)
42. Searle, J.R.: Intentionality: an Essay in the Philosophy of Mind. Cambridge University Press, Cambridge, New York (1983)
43. Copeland, B.J.: The chinese room from a logical point of view. In: Preston, J., Bishop, M. (eds.) Views into the Chinese Room: New Essays on Searle and Artificial Intelligence, pp. 109–122. Clarendon Press, Oxford (2002)
44. Leibniz, G.W.: The Monadology and Other Philosophical Writings (Trans. Latta, R.). Oxford University Press, Oxford (1898)
45. Hofstadter, D.R.: Reflections. In: Hofstadter, D.R., Dennett, D.C. (eds.) The Mind's I: Fantasies and Reflections on Self and Soul, pp. 373–382. Basic Books, New York (1981)
46. Harnad, S.: The symbol grounding problem. Physica D D **42**, 335–346 (1990)
47. Krajewski, S.: On Gödel's theorem and mechanism: inconsistency or unsoundness is unavoidable in any attempt to «Out-Gödel» the mechanist. Fund. Inform. **81**(1), 173–181 (2007)
48. Burge, T.: Individualism and the mental. Midwest Stud. Philos. **4**(1), 73–121 (1979). https://doi.org/10.1111/j.1475-4975.1979.tb00374.x

49. Fodor, J.A., Garrett, M.F., Walker, E.C., Parkes, C.H.: Against definitions. Cognition **8**(3), 263–367 (1980)
50. Cram, H.R.: Fodor's causal theory of representation. Philos. Q. **42**(166), 56–70 (2011)
51. Rescorla, M.: Cognitive maps and the language of thought. Br. J. Philos. Sci. **60**(2), 377–407 (2009). https://doi.org/10.1093/bjps/axp012
52. Alme, C.B., Miao, C., Jezek, K., Treves, A., Moser, E.I., Moser, M.-B.: Place cells in the hippocampus: eleven maps for eleven rooms. Proc. Natl. Acad. Sci. **111**(52), 18428–18435 (2014). https://doi.org/10.1073/pnas.1421056111
53. Bechtel, W.: Investigating neural representations: the tale of place cells. Synthese **193**(5), 1287–1321 (2016). https://doi.org/10.1007/s11229-014-0480-8
54. Pfeiffer, B.E., Foster, D.J.: Hippocampal place-cell sequences depict future paths to remembered goals. Nature **497**(7447), 74–79 (2013). https://doi.org/10.1038/nature12112

Chapter 4
Block's Internalist Two-Factor Account

Explaining our actions in terms of individual thoughts contributes to understanding their specificity. Let's consider the example of a boy named Johnny waiting for a bus at a bus stop. His vivid imagination allows him to momentarily escape the tedium of waiting. He imagines himself as Batman, not awaiting a bus, but Robin's magical vehicle, which they will use to embark on a quest to find the nefarious Joker. Johnny's behavior does not betray his mental engagement in a perilous mission. The child squirms, paces in circles, and may occasionally shout something. The bus arrives. Johnny, absorbed in his Batman role-play, leaps onto it as though it were Robin's motorcycle and ends up getting hurt. How can we explain this? Johnny's action is predicated on his false belief that he is Batman. The content of this belief is independent of whether Robin's motorcycle actually exists or is merely fictional. If we were to assume, in line with the principle of charity, that Johnny is aware he is boarding a bus, then we would fail to account for his injury. According to Ned Block, externalism would neglect the subjective nature of thought and fail to adequately describe situations like this one, as the content would vary depending on the existence of Robin's motorcycle. Therefore, Block advocates for a two-factor semantics, where narrow content coexists with wide content. Wide content, for instance, is supposed to explain why Johnny's action was unsuccessful—it was based on the false belief that the bus is Robin's motorcycle. However, the success of many other actions by Johnny can be explained by the fact that they were grounded in accurate representations. For example, the boy reached the bus stop because he accurately remembered its location.

The individual cognitive perspective, which narrow content accounts for, is intended to best capture the cognitive character of our thoughts. The specific role of the narrow factor is to identify the essential properties of the referents (intensionally), while the wide factor is to identify referents, regardless of how they are described (extensionally). Narrow content, responsible for the cognitive character of representations, is considered superior to wide content because, according to Block, it better addresses the needs of psychology. I will demonstrate, however, that this superiority of narrow content is difficult to maintain within a two-factor concept of

K. Bielecka, *What is Misrepresentation?*, Studies in Applied Philosophy, Epistemology and Rational Ethics 71, https://doi.org/10.1007/978-3-031-84375-4_4

content. Most importantly, I will once again challenge the notion that narrow content is content at all.

The chapter outline is as follows. First, I briefly discuss the two-factor theory of content defended by Ned Block in his paper *Advertisement for a Semantics for Psychology* [1]. After introductory remarks (Sect. 4.1), I analyze the concept of narrow content (Sect. 4.2). Then, I demonstrate the advantages and disadvantages of the theory (Sect. 4.3) and finally examine how Block's theory applies to examples from mind and brain sciences (Sect. 4.4). I defend the claim that the concept of narrow content does not lead to the naturalization of intentionality. Therefore, in the later part of the book, I argue for replacing it with the concept of wide content.

4.1 Two Factors

Ned Block proposed a certain conception of semantics for psychology aimed at reducing the semantic to the non-semantic [1: 639]. In this sense, it is a theory that should indicate what determines meaning. However, Block's intention was not to create a ready-made theory of content but to formulate a research program for such semantics in psychology, in which narrow content plays a primary role [1: 615].[1]

Narrow content first appears in the last of eight desiderata for new semantics for psychology:

> Desideratum 8: Explain why different aspects of meaning are relevant in different ways to the determination of reference and to psychological explanation. [1: 618]

In the two-factor theory, there is a narrow factor, corresponding to narrow meaning (narrow content), and a wide factor, corresponding to wide meaning (wide content)[2]:

> Nothing in my position requires me to regard narrow meaning and narrow content as (respectively) kinds of meaning and content. As mentioned earlier, I regard them as aspects of or as determinants of meaning and content. [1: 626]

However, Block does not characterize both factors unambiguously. The narrow factor defines the way in which the subject determines the essential properties of the referent, while the secondary wide factor defines the relationship between the conceptual role and the referent. It is worth noting that the nature of the wide factor is not specified in any way, and what's more—Block allows for its arbitrary definition:

> For present purposes, the exact nature of the external factor [of content - K.B.] does not matter. Those who are so inclined could suppose it to be elucidated by a causal theory of reference or by a theory of truth conditions. [1: 628]

---

[1] In this way, Block distinguishes his theory of content from Harman's theory [2].

[2] The author himself discusses narrow and wide meaning, but in essence, narrow meaning equates to narrow content, and wide meaning to wide content or reference. Below, I will use these terms interchangeably.

It is worth adding that Block extends the wide factor to include society [1: 620]. He sometimes calls the social factor the "third factor" [1: 630]. Although the referential (and social) factor plays a significant part in the explanation of behavior, in this theory, it is the cognitive factor that is decisive for the explanation of behavior.

> My highly tentative conclusion is that the aspect of conceptual role that determines the function from context to reference and truth value [narrow content - K.B.] is a small part of the conceptual roles relevant to the explanation of behavior and psychological state. [1: 670]

In other words, narrow content is meant to serve the determination of reference, or more precisely, its nature, which will be discussed in the next section.

Although Block marginalizes the role of the wide factor, he still considers it significant for explanatory purposes [1: 619, 620, 622, 623]. This distinguishes him from internalists, such as Segal or Fodor, for whom the relationship of narrow content to the nature of extension is not necessary to capture its cognitive aspect.

Block justifies the explanatory primacy of the narrow factor with two fundamental arguments concerning the semantic character of content. Firstly, the way in which we represent can influence mental states and behaviors, and secondly, semantic properties are not reducible solely to the reference of representations.

4.2 The Specificity of Narrow Content

Block identifies several features of narrow content. First, it is narrow, consistent with the assumptions of internalism (discussed in Sect. 4.3), second, it is a conceptual role, and third, it serves in determining reference or the nature of reference. This is precisely why it is content.[3] However, Block's argumentation in favor of the thesis that narrow content is indeed content is somewhat tentative. The validity of this argumentation is crucial for whether Block's concept can be called dual-factor. If narrow content were not content, it would be difficult to say that he has achieved what he intended, namely the reduction of the semantic to the non-semantic.

Block emphasizes that the narrow factor is responsible for the cognitive aspect of meaning localized 'in the head', while the wide factor accounts for the referential and social aspect of meaning 'outside the head' [1: 627]. The referential factor depends on what lies outside the cognitive system. In other words, the first factor accounts for the relations between representations 'in the head', while the second determines the function that binds these representations to reference and truth conditions, and is 'outside the head'. Terms such as 'in the head' or 'outside the head' suggest

[3] It is worth noting that Block's two-factor theory presents interpretative challenges even to psychologists. According to Susan Carey (who is Ned Block's wife), narrow content pertains to invariant cognitive factors. Therefore, all representations that are not innate cannot be considered narrow. She believes that Block suggests that various aspects of narrow content are responsible for different kinds of representational mechanisms: different representational symbols (symbols of natural kinds, symbols of logical connectors) determine different types of mapping [3]. Such an interpretation indicates the ambiguity of this concept.

that Block endorses the thesis of local supervenience on the microstructure of the cognitive system, according to which content properties supervene locally on the microstructural properties of the cognitive system, such as the physical structure of the brain. In other words, with respect to narrow content, Block endorses internalism.

4.3 Conceptual Role

Narrow content is to be the conceptual role of a given representation, where by conceptual role we mean the role that a representation plays in the cognitive life of a cognitive system, such as in perception, thinking, or decision-making [4]. At the same time, this role is defined only within the confines of the cognitive system:

> To have an internal representation with a certain narrow meaning is to have a representation with certain likely inferential antecedents and consequents. Hence, to ascribe a narrow meaning is to ascribe a syndrome of causes and effects, including, in some cases, behavioral effects (or at least impulses in motor-output neurons). [1: 668]

Block emphasizes that terms such as "inferential role" or "conceptual role" should not suggest that roles are determined in terms of meaning. The two-factor theory would then be non-reductive. That is precisely why these roles are defined in non-semantic terms [1: 631]. Block, incidentally, accepts computationalism [5], so this would be a certain role in the computations carried out by the brain.

It is worth comparing the concept of narrow content defended by Block with the concept of intensionality defined in the previous chapter as the specificity of referents. Intensionality, so understood, is not "in the head" in the case of most representations, such as the representation of a motorcycle. The specificities of the motorcycle do not supervene locally on the brain structure of Johnny from my previous example, as they are features of the motorcycle itself. It is obvious, for example, that having an engine does not supervene locally on the brain. The inconsistency of the traditional conception of intensionality with internalism could be treated as an objection to it. It would go like this: the traditional concept does not answer the question of how we recognize the features of referents, but merely assumes that somehow we do. According to the traditional conception, each concept is associated with specific features of its referents, and there are methods by which we distinguish objects possessing these features. Meanwhile, the internalist theory would have to indicate how this is possible. It would seem intuitive that in the cognitive system there must be something that accounts for recognizing the features of referents, but these are not the features of referents themselves.

One could imagine, for instance, that there are patterns in the brain used for recognizing the features of referents. However, the disposition to recognize such features does not supervene solely on the local microstructure of the cognitive system. No matter how we understand "recognition," it is always a kind of relation with the features of referents, and not all referents are located within the cognitive system itself. Meanwhile, internalism requires us to treat the disposition to recognize as a non-relational feature (that is, not having anything to do with relations with the environment). With this approach, however, it is difficult to explicate how one feature rather than another is recognized.

Block's concept offers some hope for solving this problem, as it refers to the notion of character. Block seeks an analogy between the narrow and wide factors and the concepts of character and content in the sense used by Kaplan and Perry [1: 627].[4] Character (Ch) is a function (f) from an expression in context (where the parameters of context are location, speaker, time, and possible world) to content in Kaplan's sense:

Ch = : f (context, expression) $\rightarrow$ content in Kaplan's sense,

where content in Kaplan's sense is a function mapping possible worlds into extensions, that is, the referents denoted by names, predicates, or propositions.

Narrow content is supposed to determine character in Kaplan's sense:

> Narrow meaning has another kind of theoretical import: it determines a function from expressions and contexts of utterance onto referents and truth values. [1: 621]

Block sometimes confusingly claims that character is a function whose value is reference, but elsewhere he asserts that its value is only the nature of reference:

> The two-factor theory is compatible with a variety of different mappings from a single conceptual role onto aspects of worlds. (...) What is in the head—conceptual role—determines the nature of reference without determining reference itself. [1: 644]

Of course, reference cannot be the value of this function if wide content, whose determinants are indicated by a separate theory of reference, is to be the second factor in the concept of content. In the narrow content theory defended by Fodor [6], narrow content is only a function from context and expression to reference, but this theory is single-factor, Block, however, has a weaker understanding of the relationship between narrow and wide content.

However, how do we know which aspects of conceptual roles are supposed to determine the characteristics of the referents of representations, or in other words, their specificity? Block's conception offers no answer to this. He assumes that this specifity is determined by aspects of the conceptual role, but does not specify how. In this respect, the internalist conception of intensionality does not advance beyond traditional conceptions—it presupposes intentionality instead of explaining it.

There are also doubts as to whether the use of two-dimensional semantics terminology, developed by philosophers of language such as Kaplan [7] and Stalnaker [8], actually facilitates Block's naturalization of intentionality. This semantics employs the concept of possible worlds, which were meant not to describe psychological mechanisms but to provide a rational reconstruction of semantics. The goal of two-dimensional semantics is to describe the contextual properties of language, including

---

[4] Wide content in Block's sense is not identical to content in Kaplan's sense: for Block, it is "an external component that has to do with the relations between the representations in the head (with their internal conceptual roles) and the referents and/or truth conditions of these representations in the world" [1: 627]. It is not a function, but rather the reference or logical values that pertain to a given conceptual role.

context-dependent expressions (on the utility of this approach for indexicals, see Ciecierski [9]). An opponent of Block might object that such an advanced theoretical apparatus cannot be applied to the description of basic mental representations, such as mental imagery or cognitive maps. Two-dimensional semantics can be understood weakly—as a useful tool for description, or strongly—as an account of a specific actual representational mechanism. The latter interpretation poses more difficulties: doubts about such an interpretation arise from the application of this semantics to describe not only linguistic representations but also all mental representations. The thesis that two-dimensional semantics corresponds to actual mental mechanisms requires more compelling arguments, particularly from psychological and neurological research. However, with a more limited interpretation, where two-dimensional semantics is merely a tool for description, there is no obstacle to using it to describe non-linguistic representations such as mental imagery or cognitive maps. Yet, since it does not explain intentionality in non-intentional terms, it does not advance the naturalization of intentionality by even a single step.

Susan Schneider raised a similar objection against the concept of narrow content: that there is no reason to call it "content" [10]. Schneider argues that the concept of narrow content is simply eliminated in favor of the computational role. Block himself emphasizes that the conceptual role is an inferential role, determined by the computational architecture of the cognitive system. He even concedes that it may not be recognized as a kind of meaning, but only as an "aspect or determinant of meaning" [1: 626]. The problem is that the non-semantic concept of inferential role is difficult to call an "aspect of meaning," since Block merely assumes, without providing arguments, that it determines the nature of reference. However, it is precisely the determining of the nature of reference (which is the role of character in Kaplan's sense) that should be explicated in non-semantic terms in order to achieve a reduction of the concept of intentionality.

The problem can be illustrated with an example. It is easy to identify objects in which narrow content in Block's sense occurs, without any representations. There are many physical systems in which one can distinguish inferential roles causally influencing their behaviors, and these roles supervene locally on their microstructure. This includes all functioning personal computers. Block's concept a priori grounds all symbols found in computers, as long as they carry out reasoning processes. Block cannot say that these symbols are ungrounded, because he has not narrowed down the set of features of conceptual roles that are supposed to "determine the nature of reference." They will determine the character in Kaplan's sense. This is precisely the problem, as "determining the nature of reference" remains an unresolved task in his concept: it is not enough to say that his concept is "consistent with many different mappings from one conceptual role to aspects of the world," because it is unclear which mappings are even considered and how they are determined. Block assumes that it is otherwise known which features of conceptual roles determine the features of referents and how they do so. However, this is the task of the concept of mental representations. Block himself recognizes this as an essential desideratum for any such concept: it should distinguish the content attributed to a system only from the outside (e.g., the content of sentences printed in a newspaper is attributed by the reader) from content that is meaningful in itself [1: 617]. However, since the determination of character is not naturalized, one may get the impression that this function is merely attributed from the outside.

4.4 Advantages and Disadvantages of the Two-Factor Theory

4.4.1 Two-Factor Theory of Content and Fodor's Formalist Approach

Prima facie, the two-factor theory can describe semantic differences between expressions or representations that are identical in form but differ semantically, distinguishing it from Fodor's formalist concept. Fodor's method of differentiating content based on form is effective where the semantic difference is determined by form. Block, not applying a formal criterion in the style of Fodor, can nevertheless describe this difference in terms of narrow content. It seems he could articulate what differentiates sentences like "I have big feet" used in various contexts. In different places, the same expression can have different wide content.

Block's concept of concept learning is similar to Fodor's. He accepts Fodor's basic assumption that learning is based on hypothesis formation and testing, but rejects his innateness hypothesis. He unequivocally dismisses the claim that innate concepts include ZIPPER, ENZYME, TRANSISTOR, MESON, and MONITOR as evidently false [1: 637]. Yet, he explains concept learning differently, settling for a functional description where acquiring concepts is treated analogously to language acquisition.

4.4.2 Narrow Content in the Dual-Factor Concept and Behaviors

Block uses the analogy of narrow content to character in Kaplan's sense to point out a certain advantage of narrow content over wide content in causal explanation [1: 619]. I will illustrate the difference between two aspects of meaning with an example analogous to the one proposed by Block.

Suppose we describe the behavior of John Smith, who delusionally believes he is Napoleon. Compare the sentences:

(1) I am Napoleon.
(2) John Smith is Napoleon.

Sentence (2) is causally irrelevant for explaining what John Smith, who considers himself not as John Smith but as Napoleon, means. The pronoun "I," as Block argues, best expresses the individual character of content.

According to Block, content can be distinguished independently of the context in which it appears, especially independently of the subject of the proposition. The above example shows that narrow content expressed through a conceptual role is useful for psychologists in grouping, and more widely classifying behaviors, such as pathological behaviors like delusions, but also healthy behaviors like thinking in terms of fiction.

In summary: Block argues that both narrow and wide content play a role in explaining and predicting individual behavior. Wide content explains the behavior of a specific individual, while narrow content provides greater generality in explaining behaviors independent of specific persons.

4.4.3 Objections to Block's Dual-Factor Concept

Fodor and Lepore's Objection: Uncoordinated Factors

The primary problem with the two-factor theory is that the narrow and wide factors diverge [11, 12]. According to Fodor [11], what are mutually uncoordinated are the inferential factor, which determines the relations occurring between propositions, and the referential factor, which is responsible for relations with the world. For example, the proposition 'Water is wet' has two different conditions of satisfaction and is connected (1) inferentially with a proposition about water and (2) with H_2O. Fodor believes that in the two-factor theory, it is unclear which exact narrow content is associated with which wide content, which may lead to the conclusion that the two-factor theory is not actually one theory, but rather an amalgamation of two distinct conceptions.

In relation to Block's conception, this criticism appears to be well-founded. Block characterizes narrow content as a conceptual role, yet fails to demonstrate which aspect of this role would determine the nature of reference and the manner in which it would do so. Furthermore, his wide content is not fully defined by the narrow, because Block emphasizes that aspects of the conceptual role determine the nature of reference, not the reference itself. Since he distinguishes the nature of reference from the reference itself, it implies that there is some difference between these concepts, and the connection of this nature with the reference itself remains ambiguous. Block asserts that his concept is compatible with various mappings from the conceptual role to aspects of the world. Suppose that within the cognitive system there exists the representation WATER, whose conceptual role—as Block contends—is defined by all the inferences in which it is involved. He would likely argue that in different worlds, where either H_2O or XYZ exists, this role would identify one of them, and in a world where both substances coexist, the representation WATER could not serve to distinguish between them. However, Block provides no rationale to believe that the computational role of the representation WATER is in any way connected to its referents. He merely assumes that the conceptual role determines character in Kaplan's sense. Yet, this assumption remains unargued. Block presupposes what he was supposed to demonstrate, namely, that conceptual roles are related to content. His assumption thus leads to circular reasoning and is therefore highly controversial. Even if we were to accept that this assumption is valid, Block would still need to show that the function in the nature of reference has values consistent with what some theory of reference would attribute to this representation. Block himself concedes that the choice of a wide content theory is quite arbitrary, and he refers to the causal theory of reference and the theory of truth conditions. However, according to Putnam's causal theory of reference, natural kind terms are rigid designators and cannot refer to XYZ.[5] In a world devoid of H_2O, the narrow content would indicate the nature of XYZ, and according to the causal theory of reference, this representation would be vacuous. In this respect, these factors are entirely uncoordinated.

In the previous chapter, I argued that an adequate concept of mental representation should incorporate reference. Although Block's two-factor theory does consider it, it fails to ensure that intension and extension will be properly connected.

In summary, considering that Block characterizes wide content so liberally without specifying what determines it, it remains unclear whether two different procedures for determining reference—one associated with the function from context to reference, and the other with a separate concept of wide content—would not be in conflict. According to Fodor, this consequence is the price of adopting a dual-factor conception, because in principle these factors can be determined by different methods and thus designate different references if they are indeed independent. Otherwise, we would be dealing with aspects of the same factor, and the threat of non-coordination would not arise.

Fodor's Objection Against Inferentialism

Fodor presented many other critical arguments against inferentialism. He believed that in Block's conception, narrow content entangled itself in problems with the compositionality of concepts: inferentialism does not guarantee compositionality.[6] However, contrary to Fodor, one might argue that it does not necessarily exclude it. One of the more frequently cited arguments is that of Fodor and Lepore, who demonstrate that inferentialism assumes holism [14, 15].[7] As the authors of the objection claim, inferentialism leads to the following dilemma: we must either accept (a) that differences in opinion lead to differences in meaning, or (b) what they consider to be the implausible distinction between analytic and synthetic statements [16]. Such a strong assumption that every difference in opinion leads to a difference in meaning would make understanding impossible, as substantive elements of disagreement can arise when interlocutors share at least part of the language and its meaning [17]. This would contradict the principle of charity in its limited version, which I defend (see

[5] The assumptions of the causal theory are presented in the next chapter.

[6] The difficulties in developing compositional inferentialism are quite significant, although Brandom [13] believes they are surmountable.

[7] Semantic holism claims that the meaning of individual sentences (representations) should be understood in terms of relations with other sentences occurring in the language (other representations in the system).

Chap. 2). The second assumption is unacceptable for reasons given by Quine [16].[8] Since adopting either solution is undesirable, Fodor and Lepore argue, we must reject both holism and inferentialism.

The above criticism of inferentialism is unsatisfactory because it addresses a caricature of the theory, according to which every inference equally constitutes meaning. Identifying only those inferences that constitute meaning, thus defining holism, is a concrete challenge that any serious inferential theory must face. Fodor and Lepore's arguments are inconclusive for both inferentialism and holism. Holism of representation and the structure of the representation system are tied to having a structure of the representation system. Since all representations have a more or less rich structure, it is difficult to imagine representations that are not holistic in a minimal sense, in that they could not be used in inferences. Inferences, in turn, determine relations of inferential equivalence, etc.—perhaps some of them are semantically significant. Fodor and Lepore are correct, however, that it is necessary to clearly indicate which ones. Block indeed sees this as a task for the inferentialist conception and suggests that it would replace the dichotomous distinctions between analytic and synthetic with a matter of degree:

> Without an analytic/synthetic distinction, we would, as I mentioned earlier, have to move to a scientific conception of meaning that does away with the crude dichotomy of same/different meaning in favor of a multidimensional gradient of similarity of meaning. [1: 629]

It is worth noting that Block's views are close to those of Quine, who does not reject this distinction categorically but considers it to be gradable rather than dichotomous.

Perlman's Objection

Against the two-factor theory, a similar argument was put forward by Perlman [23]. He believes that misrepresentation is impossible within its framework because the semantics of conceptual roles as such does not allow for the description of its incorrectness. Conceptual roles are abstracted from regularities in inferences and other uses of concepts, but since one can err in inference and also misapply a concept therein, such errors would also become the 'norm' in light of this semantics. According to him, solving this problem requires the distinction between analytic and synthetic propositions, as otherwise, it is impossible to differentiate inferences constitutive of meaning from inferences that are irrelevant.

Perlman, however, underestimates the difficulties associated with defining the erroneousness of representations. Even assuming that Block could devise some

---

[8] As Barbara Stanosz observes, Quine articulated four arguments against adopting this distinction: "(1) reference to the concept of definition does not lead to an explanation of the concept of synonymy, as defining assumes the relation of synonymy as previously given; (2) substitutibility in extensional contexts is not an adequate criterion of synonymy, while interchangeability in intensional contexts can only be established given a prior concept of analyticity; (3) the concept of a semantic rule, which definitions of analyticity in artificial languages rely on, is incomprehensible; (4) the verificationist theory of meaning indeed allows for the determination of synonymy, but it presupposes reductionism—the second dogma of empiricism" [16]. However, Quine contends that analyticity may apply to propositions to varying degrees, depending on the extent to which they are subject to the "tribunal of experience.".

multidimensional criterion of meaning similarity, it is far from clear that he could use it to identify erroneous representations or inferences. The concept of mental representations should also explicate the very notions of error or correctness in a naturalistic manner, which are considered normative concepts in at least a minimal sense. The concept of conceptual role itself is defined in terms of being completely non-normative. Block responds to the criticism of neglecting error by claiming that perhaps this problem can be resolved by referring to the disposition of cognitive systems to correct their errors, but he does not show how this is to be integrated into his theory [4]. In subsequent chapters, I defend the claim that the problem of error can only be resolved within teleosemantics.

In summary, the most pressing issue of the relation between narrow content and wide content remains unresolved. This is related to the relation between inferences characterized causally and reference. Fodor and Lepore's criticism has not been rebutted by Block. His proposal for solving the problem of erroneousness is unsatisfactory.

4.5 Application of the Account to Examples from Mind and Brain Sciences

Rombix

I will illustrate the role of narrow content in psychology with the example of "rombix." The narrow content of the concept "rombix" is determined by inferences conceived in individualistic and computational terms. It remains the same regardless of context (or possible world). The wide content of the concept "rombix" is established, for example, by the causal theory of reference as a red diamond-shaped spot on a square background.

Statements of similar form, such as: "A rombix is a diamond," uttered by twins on different Earths may have the same conceptual roles without having the same reference. In the second world, for instance, there may be an inverted color spectrum, and then "red diamond" would mean a green diamond (from our perspective). A person referring to an Earthly rombix has the same narrow content as a corresponding person in a counterfactual possible world where rombices are green. Wide content, however, allows us to assert the difference in reference, as propositions on both Earths may refer to objects with inverted colors.

The content of the concept "rombix" is to be determined by its conceptual roles. As the subject gradually becomes familiar with this concept by observing various examples and counterexamples, their initial conception of a rombix may be erroneous, which the experimenter will indicate. Unfortunately, Block's semantics of conceptual roles is non-normative, thus it fails to provide clear criteria for correct usage. As I demonstrated in the previous section, Perlman argues that it merely describes certain regularities in the use of representations, without offering criteria for distinguishing correct from incorrect uses. Conditions for the identity of conceptual roles, though not specified by Block's conception—who reserves that this is a matter for further development of the theory [1: 624]—are certainly individualistic. This is because conceptual roles supervene locally on the microstructure of the cognitive system.

Let's assume that the subject claims that a rombix is a red peg. Since they employ the concept ROMBIX and it figures in their reasoning, according to the semantics of conceptual roles, the subject possesses this concept. From their perspective, it is the experimenter who is mistaken. The two-factor theory allows for an appeal to the social aspect of meaning, but it is unclear how this would be helpful in this situation. One cannot presume that concepts are innate, as Block rejects Fodor's nativism. Since his concept of conceptual roles remains only nominally related to the wide or social aspect of meaning—as Fodor and Lepore point out—Block cannot assume that the subject and the experimenter share a set of elementary concepts that would form the basis of understanding. It is always possible that the same conceptual role (narrow content) will coexist with a completely different wide content. Block considers this a strength of his theory, as it allows for the description of the invariance of narrow content amidst changes in wide content (as in the Twin Earth experiments). The crux of the matter is that even in the same situation, two interlocutors need not share any wide content at all, and in the case of misuse of terms by one of them, one cannot determine who is speaking correctly and who is not. Meanwhile, in the experiment, the subject concedes that the experimenter is right and corrects their use of the concept.

Block's theory fails to explain or predict the subject's behavior because, according to it, there would be no way to convince the subject that they are mistaken. This is because, in Block's concept, various factors of meaning (social and narrow) can be uncoordinated. Without a normative perspective (typical of other types of inferentialism), it is difficult to find rational arguments that aid in intersubjective understanding.

Purely descriptive inferential semantics does not account for the normativity of representations, a key aspect of concept learning. Therefore, it is unsuitable for a complete description of at least some psychological experiments in which subjects make mistakes—and are corrected.

In Block's work, as in Stich's, there arises the problem of reducing meaning to form, as conceptual roles are distinguished as purely computational. Block claims, but without justification, that these roles determine the function from context and conceptual role to the nature of reference; in other words, he believes that these roles confer intentionality to representations. Since it is unclear how they would be intentional, an opponent of Block could argue, as Schneider does, that in the rombix experiment, the two-factor theory provides nothing beyond a description of the computational system's functioning. With such an interpretation, it becomes even more difficult to speak of the correctness or incorrectness of the learned concept ROMBIX. Therefore, due to the unsuccessful reduction of intentionality to computational concepts, the two-factor theory is not a good tool for describing experiments on the acquisition of artificial concepts.

Cognitive Map

In an experiment conducted in 1932, a rat learned the layout of the maze's arms, even though there was no food, the reward element, present. However, when food was placed in the maze, the rat immediately found it, even though it was not hungry [18]. In this case, the rat represents the path correctly and reaches the place where food had previously been found. However, the rat reacted with surprise when it traversed the same path that had led to food before, but this time there was no food. In this instance, it represented the path incorrectly.

Block might say that there are computational roles in animals that are associated with processes of recalling, predicting, planning, movement, etc. In such terms, he might also try to describe the functioning of the cognitive map as narrow content. An opponent of the two-factor theory might criticize Block for characterizing content through language, thereby rendering his theory unsuitable for describing animal representations, especially structural ones. However, it seems that the two-factor theory a priori does not exclude non-linguistic representations.

Although Block primarily writes about concepts, using the notions of propositions and characterizing content by its inferential roles, his notion of narrow content does not exclude other representations if one considers inferential roles in computational terms. The thesis that thinking consists of inference—and in accordance with the rules of logic—might seem to some as an overly intellectualized view. However, Block need not understand inference in this way; the concept of unconscious reasoning has a long tradition in psychology [19]. It provoked the introspectionists' objections but was rehabilitated by cognitive science in studies on grammar and perception. One of the modern concepts that attribute a central role to inference is the Bayesian brain hypothesis, which posits that thinking is based on probabilistic inference (according to Bayes' rule). In the approach of Andy Clark and Karl Friston, such inferences are conducted hierarchically by the nervous system, brains "predict" future sensory experiences through top-down predictive processing and minimize the errors of these predictions in two ways: by changing the predictions or by acting upon the world [20, 21]. Inferentialism may refer to concepts that assume not so much logical reasoning, but psychological processes of reasoning—this is suggested by Harman, who argues that inferentialism would benefit from an empirical theory of inference grounded in psychology, not logic [17]. Another matter is that, to date, this theory has not been developed in this direction.

Thus, one could attempt to describe the functioning of cognitive maps in terms of computational roles and reasoning, such as probabilistic inference. This kind of approach is consistent with the practices of modern computational neuroscience.

Block might say that the same narrow content is possessed by a cognitive map in both mazes. The theory would predict that the rat will exhibit the same behavior until it becomes aware that there is no food in the location it seeks in the second maze. This is because the rat has the same narrow content.

Given the non-normative nature of Block's concept, one might question whether it accurately describes the misrepresentation of a rat when it turns out that the expectation of food being in a certain place in the maze is unmet. Nevertheless, in contrast to methodological solipsism, Block's conception does not preclude speaking about the truth or falsity of representations within psychology itself, thus does not exclude speaking about the conditions of satisfaction. It is particularly evident that reference can play a role in Block's view when a rat is transferred from one maze to another. The rat's surprise in a twin maze remained an unexplained anomaly in light of methodological solipsism, whereas Block could assert that the rat realized it was in a different maze than before. This occurs only by virtue of a causal (or other) theory of reference, which is a wide factor in his concept. The narrow content, however, remains the same.

Mental Imagery

If the arguments concerning the possibility of including cognitive maps in the semantics of conceptual roles are correct, then they apply even more so to mental imagery. Assuming that this semantics is not limited to the language of thought, it also copes with the second example of structural representations, the example of mental imagery. Block himself defended the non-linguistic nature of imagery [22]. Here, the advantage of Block's conception over Fodor's internalism is evident.

Mental imagery appears when they are not perceptions. Block might emphasize that this is why they are better characterized in terms of narrow content, because in this case, the reference is always empty: the images represent something that is not currently present. Wide content would be useful when one needs to compare mental images with currently seen drawings, to determine whether two illustrations of the same figure represent the same figure or different figures, as in the drawing below by Ernst Mach. In the case of pair A, one almost immediately notices that they are the same figures, while in the case of pair B, proponents of mental imagery believe that one can imagine rotating the right figure and compare it with the left (Fig. 4.1).

In this experiment, the mental imagery of the right figure is compared with the current perception of the left figure. Block encounters a problem in determining how to compare narrow content with wide content—it is only the latter that enables the comparison of mental imagery with the actual drawing. Therefore, in the current version of the two-factor theory, one can only harbor the hope that such experiments involving mental imagery will be successfully described.

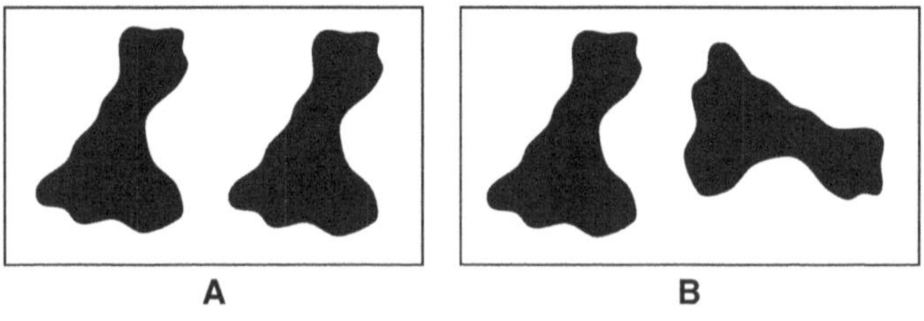

Fig. 4.1 Two pairs of perceptual stimuli to be associated in mental imagery (Fig. Ernst Mach)

4.6 Summary

Block's theory incorporates the wide content factor, thereby avoiding some of the problems of single-factor theories. It can, for instance, accurately describe experiments involving transferring rodents from one maze to another, and it does not become entangled in the difficulties associated with defending formalism or nativism, characteristic of proponents of methodological solipsism.

Block demonstrates that the explanatory role of both factors is different, yet he fails to show that they are interconnected. As I have argued, pluralism in the concept of content is unsustainable due to the difficulties in coordinating the two factors. Block at best presupposes intentionality, rather than explaining it. Narrow content does not seem intentional at all. Therefore, it cannot be said that Block's theory naturalizes intentionality. It should be noted that this is not an accidental feature of two-factor semantics, but of all internalist theories that employ the concept of narrow content. The intentionality of narrow content must, by definition, supervene locally on the microstructure of the cognitive system.

In response, an internalist might argue that narrow content need not be intentional. In such a case, however, it is difficult to justify why we should call it "content" when there is a better and simpler concept: causes of behavior and cognitive processes (or the computational role, as in Block's case). As I pointed out in Chap. 2, intentionality is a paradigmatic feature of mental representations, and only those features of representations that exhibit intentionality should be termed "content."

Conceptual roles are supposed to determine the nature of reference, but there is not even an outline of a proposal for how this would be accomplished. Contrary to Block's intentions, the two-factor theory is not reductive, and at best, it is halfway between an intentional stance and a functional stance. Contrary to Block, the content in the two-factor theory seems to be attributed to conceptual roles solely based on the interpreter's knowledge, rather than the structure of the cognitive systems themselves.

The two-factor theory of content fails to meet desideratum II.2 on the naturalization of intentionality. I have shown that the concept of narrow content does not allow it to be explicated; Block does not explain how the concept of wide content would serve this purpose. He also does not clarify why truth is an essential element of the philophical account of mental representation, thus failing to meet desideratum II.2.2 on the bearer of content. The two-factor theory also does not fulfill desideratum II.3 on the proper account of content, primarily because it does not adequately capture the relation between intension and extension, since reference and content can remain uncoordinated. Block does not analyze the problem of the possibility of error; since the theory of narrow content does not allow for the adjudication of incorrectness (as I demonstrated in Chap. 3), only the conception of wide content could potentially address it. However, Block's treatment of wide content remains cursory and schematic, leaving it an open question whether it would allow for the discussion of misrepresentation.

According to Block, desideratum II.3.1 on productivity is satisfied. This is supposedly because the two-factor theory remains compositional, unlike other semantics of conceptual roles [4]. However, the claim that it is compositional is overstated: it has not been elaborated in detail, and attempts to incorporate compositionality into inferentialism indicate that, contrary to Block's assurances, this is not at all straightforward [13].

In this part of the book, I have analyzed various naturalistic theories of narrow content. The reasons of proponents of internalism, such as the cognitive independence of thought from context, cannot be explicated by the conception of narrow content—neither monistic nor two-factor. As I have argued, the cognitive character of content cannot be captured in a conception that does not demonstrate the contentfulness of narrow content and how the wide factor is determined.

In summary: internalist arguments concerning the cognitive character of narrow content are insufficient to defend either the monistic theory or the two-factor theory, which treats the narrow factor as primary over the wide factor. Therefore, I reject internalism in favor of externalist theories, in which wide content forms the basis of semantics and can satisfy certain internalist intuitions, especially those related to the requirement that content should be understood not only from an external perspective but also within the dynamics of the cognitive system itself. In the subsequent part of the work, I limit myself to the analysis of positions on wide content.

References

1. Block, N.: Advertisement for a semantics for psychology. Midwest Stud. Philos. **10**(1), 615–678 (1986). https://doi.org/10.1111/j.1475-4975.1987.tb00558.x
2. Harman, G.: (Nonsolipsistic) conceptual role semantics. In: LePore, E. (ed.) New Directions in Semantics, pp. 55–81. Academic Press, London (1987)
3. Carey, S.: The Origin of Concepts. Oxford University Press, Oxford, New York (2009)
4. Block, N.: Conceptual role semantics. In: Craig, E. (ed.) Routledge Encyclopedia of Philosophy, pp. 242–256. London; New York (1998)
5. Block, N., Fodor, J.A.: What psychological states are not. Philos. Rev. **81**(2), 159–181 (1972)
6. Fodor, J.A.: Psychosemantics: the Problem of Meaning in the Philosophy of Mind. MIT Press, Cambridge (1987)
7. Kaplan, D.: On the logic of demonstratives. J. Philos. Log. **8**(1), 81–98 (1979). https://doi.org/10.1007/BF00258420
8. Stalnaker, R.: Context and Content: Essays on Intentionality in Speech and Thought. Oxford University Press, Oxford, New York (1999)
9. Ciecierski, T.: Zależność kontekstowa. Wprowadzenie do problematyki. Wydawnictwo Naukowe, Warszawa (2011)
10. Schneider, S.: The Language of Thought: a New Philosophical Direction. MIT Press, Cambridge (2011)
11. Fodor, J.A.: Why paramecia don't have mental representations. Midwest Stud. Philos. **10**(1), 3–23 (1986). https://doi.org/10.1111/j.1475-4975.1987.tb00532.x
12. Perlman, M.: The trouble with two-factor conceptual role theories. Mind. Mach. **7**(4), 495–513 (2004)
13. Brandom, R.: Between Saying and Doing: Towards an Analytic Pragmatism. Oxford University Press, Oxford (2010)
14. Fodor, J.A., Lepore, E.: Why meaning (probably) isn't conceptual role. Mind Lang. **6**(4), 328–343 (1992)

15. Fodor, J.A., Lepore, E.: Holism. A Shopper's Guide. Blackwell, Oxford (1993)
16. Quine, W.V.O.: Two dogmas of empiricism. Philos. Rev. **60**(1), 20–43 (1951). https://doi.org/10.2307/2181906
17. Greenberg, M., Harman, G.: Conceptual role semantics. In: Lepore, E., Smith, B. (eds.) The Oxford Handbook of Philosophy of Language, pp. 295–322. Oxford University Press, Oxford (2006)
18. Tolman, E.C., Honzik, C.H.: Introduction and removal of reward, and maze performance in rats. Univ. Calif. Publ. Psychol. **4**, 257–275 (1930)
19. Helmholtz, H.: Handbuch der physiologischen Optik. Leopold Voss, Leipzig (1867)
20. Clark, A.: Whatever next? Predictive brains, situated agents, and the future of cognitive science. Behav. Brain Sci.. Brain Sci. **36**(3), 181–204 (2013). https://doi.org/10.1017/S0140525X12000477
21. Clark, A.: Surfing Uncertainty: Prediction, Action, and the Embodied Mind. Oxford University Press (2016)
22. Block, N.: The photographic fallacy in the debate about mental imagery. Noûs **17**(4), 651–661 (1983). https://doi.org/10.2307/2215087
23. Perlman, M.: The trouble with two-factor conceptual role theories. Minds & Machines 7(4), 495–513 (1997)

Chapter 5
Externalist Conceptions of Representation

In this (and the following) chapter, I examine non-teleosemantic externalist conceptions. I investigate whether the concept of reference or truth conditions alone can supplant the concept of content and simultaneously naturalize intentionality. In this chapter, I analyze causal-informational semantics, which is confined to the concept of reference, while in Chap. 6, I turn my attention to Donald Davidson's truth-conditional account.

The structure of this chapter is as follows: the chapter is divided into two main parts, the first concerning causal semantics (5.1), and the second—Jerry Fodor's asymmetric conception (5.2).

5.1 Informational-Causal Semantics

After presenting preliminary remarks (Sect. 5.1.1), I briefly introduce informational-causal Wisconsin-style semantics (Sect. 5.1.2). I then discuss the problems it leads to: the problem of reference determination (Sect. 5.1.3), which includes the disjunction problem (Sect. 5.1.4), and the problem of trivialization (Sect. 5.1.5). I present some solutions to the reference determination problem (Sect. 5.1.6), and finally, I examine how the discussed view can describe examples from mind and brain sciences (Sect. 5.1.7).

K. Bielecka, *What is Misrepresentation?*, Studies in Applied Philosophy, Epistemology and Rational Ethics 71, https://doi.org/10.1007/978-3-031-84375-4_5

5.1.1 Introduction to Informational-Causal Semantics. Motivations

Kim [1] distinguishes two types of causal semantics: (1) Wisconsin-style semantics [2–4],[1] according to which reference is characterized informationally and causally (hence this semantics is also referred to as "informational-causal semantics"), and (2) Kripke-Putnam semantics [5, 6], according to which reference is characterized by causal-social connections. Given the fundamental task of the chapter, formulated as the question: Can and how can reference be determined in non-intentional terms, I confine myself to discussing informational-causal semantics.[2]

This semantics employs the concept of natural meaning. It is a property of representations, constituted by correlation or causality that occurs by virtue of laws or probable dependencies. Natural meaning can be defined as follows:

> Sign S naturally signifies C if and only when C is the cause of (or co-varies with) S.

Representations have natural meaning if and only if they are effects (or correlates) of their referents; if the referents do not exist, then we cannot speak of natural meaning. Representations either refer to their referents or are not representations at all. This leads to undesirable consequences for a conception that aims to account for the possibility of misrepresentation, which will be discussed later in the chapter.

Natural meaning, constituted causally or through co-variance, can also be defined as "indicating."[3] Examples of indicating are provided by simple artificial devices, such as a thermometer, as well as biological systems, like sensory receptors or neuroreceptors. For instance, if a thermometer indicates a temperature of 38 °C, then the external temperature is precisely such that it causes this particular indication on the thermometer. If the external temperature is different, then the thermometer will not indicate 38 °C (assuming the thermometer is not broken). The functioning of a frog's retina can also be described in terms of natural signs [7, 8]. A natural sign on the frog's retina is a group of neurons that activate when a moving insect appears on the horizon.

---

[1] The idea of naturalizing intentionality in terms of causality and correlation was developed by Fred Dretske and Dennis Stampe at the University of Wisconsin. Hence, inspired by Fodor's terminology [3], I refer to this conception as 'Wisconsin-style semantics'.

[2] The fundamental task of semantics in the Kripke-Putnam framework is not the naturalization of reference; it is the description of the social practice of language use. In this book, I utilize certain elements of this practice's description, but I treat them more as a supplement to the main task of the chapter. I thus abstract from both the concept of intension and from the capacities, knowledge, or intentions of the cognitive system, which Kripke-Putnam semantics sometimes employs to explain this practice. For similar reasons, I also exclude the hybrid version of the causal theory of reference, in which informational-causal semantics is combined with descriptivism.

[3] Natural meaning is sometimes conceived in terms of semantic information [2], defined by a correlation based on probabilistic relations. In semiotics, the concept of natural meaning can also be defined through similarity (usually structural) or mapping. However, Wisconsin-style semantics does not make use of meaning as defined in this way.

5.1.2 *Establishing Reference and the Problem of Reference Determination*

In informational-causal semantics, it is causality or covariation between a referent and a property of the cognitive system that constitutes mental representations.[4] The problem is that it is not known what the potential referent of a given representation is and which features of that referent are significant—this is known as the problem of reference determination [9]. This problem pertains only to externalist conceptions of representation, as they utilize the concept of representation's reference—unlike internalist conceptions. This issue has been widely discussed, and a particularly challenging version of it has been identified as the disjunction problem [10, 11]. Causal and correlational relations alone are insufficient to determine reference, leading to the disjunction problem (the essence of this problem is presented in the next section). In other words, the concept of representation used in this conception seems too broad.

5.1.3 *The Disjunction Problem*

The disjunction problem is one of the significant issues in the determination of reference [10, 11]. It arises when it is unclear whether the representation is caused by A and only A, or by the disjunction 'A or B'. Imagine a drunken Vlad who calls water "vodka." The water is in a bottle labeled "Wyborowa." According to the informational-causal view, what determines the reference is the connection (causal or correlational) between Vlad's mental state and the referents of his representations. Meanwhile, Vlad joyfully exclaims "Vodka!" at the sight of a bottle of water. What, in the current situation, is the extension of the term "vodka"? Perhaps Vlad is mistaken, or perhaps he is using a peculiar idiolect. Prima facie, there are three possible answers:

(a) "vodka" refers to vodka (in this case, Vlad is mistaken);
(b) "vodka" refers to water or vodka (the liquid being water or vodka);
(c) "vodka" refers to water.

The problem is that case a) cannot be described using the tools of informational-causal semantics. The choice is between option c), a bizarre case that would need to be eliminated if Vlad shouted "Water!" at the sight of water in a lake, or b), the peculiar idiolect of Vlad, in which "vodka" means a liquid that is water or vodka. The disjunction is unavoidable in the case of any error in naming referents. For the denotation of a concept includes every named object or its property, even when the object has been mistakenly named.

The essence of the disjunction problem is the lack of a criterion that would decide what belongs to the extension of a given term and what does not. I distinguish two types of the disjunction problem, which I call, following Peter Godfrey-Smith [12],

---

[4] The concepts of constitution and determination are equivalent in naturalistic conceptions, as they concern establishing the essential causes or correlations for being a representation.

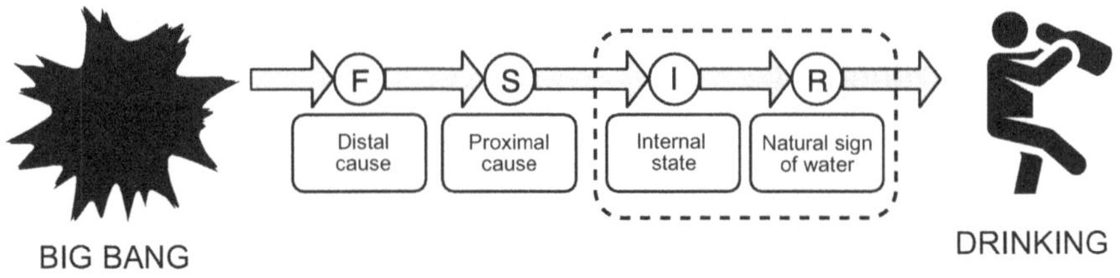

Fig. 5.1 Horizontal disjunction problem

"horizontal" and "vertical," respectively. The horizontal disjunction problem (see Fig. 5.1) concerns the distality of the cause of a given representation.

In the informational-causal theory of reference, it is unclear what constitutes the referent, as causal chains can potentially extend far into the past—from the nearest cause (the most proximal; marked as "S" in the figure) to the farthest (the most distal; marked as "F" in the figure). Thus, anything that is an element of a given causal chain can be a natural sign (marked as "R"). WATER could therefore denote an internal state, sensory stimuli elicited by a watery liquid, H_2O, the Big Bang… The problem is manageable in that it pertains to any causal explanation, and any theory of causation that distinguishes relevant causes for a given phenomenon from background causes can resolve it. Such a solution is provided by John Mackie's theory of causation [13] or James Woodward's interventionist theory [14].[5] Therefore, this is not a problem specific to informational-causal semantics. The vertical disjunction problem (see Fig. 5.2) boils down to the question of which of the possible causal factors at a given stage of the causal chain is the referent.

This problem is challenging because identifying the referent is not sufficient for its resolution. The Evening Star is, after all, the Morning Star, as both refer to the planet Venus. However, the name "Evening Star" has a different intension than "Morning Star," because we use the former to refer to the planet Venus visible in the evening, and the latter to refer to the planet Venus visible in the morning [15]. The crux of the matter is that a causal (or correlational) relation per se does not distinguish these properties. The problem is not hopeless, though, because visibility in the morning and visibility in the evening have different consequences, informational-causal semantics should thus be able to exploit this fact. Unfortunately, it seems that without additional assumptions, Wisconsin-style semantics does not cope with this problem. Someone's use of the name "Evening Star" would correlate with a denotation that is the union of the denotations "Evening Star" and "Morning Star" (described by the disjunction "Evening Star $\vee$ Morning Star").

The example of Vlad illustrates the issue of identifying the essential properties of a referent. When the symbol VODKA is triggered, its extension automatically expands to include a new property. If Vlad's representation of vodka is caused by a

[5] I do not examine the details of these solutions. For the purposes of the discussion with informational-causal semantics, I assume that some version of the concept of causal explanation is maintainable, and the theories I point out align very well with the practice of explanation in special sciences, making them particularly promising against other accounts of causal explanation.

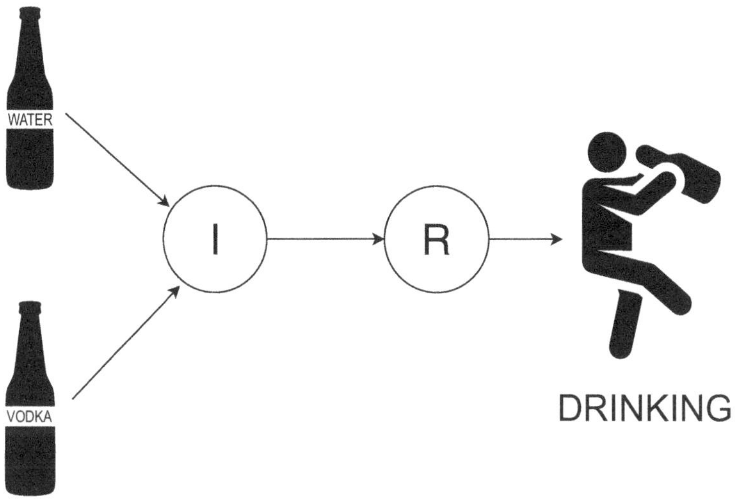

Fig. 5.2 The vertical disjunction problem. The labels in Fig. 5.2 correspond to those in Fig. 5.1

different property than previously (e.g., when this representation, typically caused by vodka, is triggered by a bottle of water), then its extension automatically expands to include a new property, taking the form of a disjunction (a liquid being either water or vodka). Consequently, the extension of what the neural state symbolizes includes not only what has caused the representation in the past but also what it refers to now. Thus, without additional criteria for distinguishing the properties of referents, informational-causal semantics fails to address this problem.

Moreover, causal or correlational relationships alone do not allow us to differentiate misrepresentations from the absence of representation, making it impossible to indicate the appropriate referents of a given representation—which is the essence of the disjunction problem. Since the disjunction problem arises, one might doubt whether there are satisfaction conditions of representations: if all representations can be considered correct, then it is impossible to determine their satisfaction conditions. Consequently, such an informational-causal account cannot naturalize truth.

Although the informational-causal view is naturalistic and—unlike internalism—takes reference into account, it does not fully meet many of the desiderata outlined in the introduction. Thus, the desideratum of determining reference and content (II.2.1) is not completely met, as this conception struggles with the determination of content, nor does it meet the desideratum of the content vehicle (II.2.2) and both desiderata concerning error (II.3.2 and II.3.3).

5.1.4 The Trivialization Problem

The trivialization problem arises when a representation appears as a result of any causal factor, thereby trivializing the concept of mental representation, making it coextensive with the term 'effect'. It is also insufficient to limit the set of causes to those that provoke a given behavior. For example, if a dog eats a newspaper, it becomes a cause of the dog's behavior, yet we would not consider it a representation of the dog. Thus, informational-causal semantics does not meet Ramsey's job description challenge [16], which is to indicate a role specific only to mental representations. In informational-causal conceptions, where representational causal roles are not distinguished from others, the concept of representation becomes trivialized, as any effect may turn out to be a representation of its cause [16].

5.1.5 Solutions to the Problem of Reference Determination. Normal Conditions

The concept of normal conditions has its tradition as a criterion for distinguishing correct from incorrect representations [11, 17]. The proposed solution is as follows: under normal conditions, representations refer correctly, while under other conditions, they do not.

The easiest way to define such conditions would be in statistical terms. After all, Vlad does not always confuse water with vodka; it is probably quite rare. Numerous examples of misrepresentations appear in biology. Ruth Millikan cites the behavior of beavers, which, suspecting the presence of a predator, immediately sound an alarm by slapping their tails against the surface of the lake [28: 288]. The thing is that most of these alarms are false. Better to make noise and flee than to be eaten. Frequency, in short, does not pass the test as a criterion for defining normal conditions. In the case of beavers, the most frequent are indeed misrepresentations.

Therefore, most definitions of normal conditions rather refer to counterfactual conditional periods. Nonetheless, as Fodor shows, the very idea is doomed to failure, as it requires the differentiation of two types of situations: type 1, where normal conditions occur, from type 2 situations. In type 1 situations, we can never be mistaken, while in type 2 situations, the possibility of error exists. Fodor defines a type 1 situation as follows:

(i) If it's a law that *P*s cause *S*-tokens in type one situations, then *S* means *P* (and if *P* is disjunctive, then so be it);
and
(ii) not all situations in which *S* gets tokened qualify as type one, so that tokens of *S* that happen in *other* sorts of situations are ipso facto free to be false [11: 64].

Thus, symbols either (A) always represent truly under ideal conditions or (B) may misrepresent under non-ideal conditions. According to Fodor, however, the distinction between the two types is untenable.

> The basic idea is to distinguish between two types of situations, such that lawful covariation determines meaning in one type of situation but not in the other. The revised theory says, in effect, that a symbol expresses a property if instantiations of the property are nomically sufficient for instantiations of the symbol *in situations of type one*. Since the tokens of a symbol that occur in type one situations are ipso facto caused by things that are in its extension, it follows that all such tokens are true. However, properties whose instantiations cause tokens of a symbol (only) *in situations of the second type* are *not* thereby expressed by the symbol; so tokens of a symbol that occur in type two situations are *not* ipso facto caused by things in its extension; so it is left open that such tokens may be false. [11: 60]

In type 1 situations, normal conditions prevail, described in terms of probable dependencies and counterfactual conditional statements:

(1) Symbol S (instances of symbol S) expresses property P if the generalization "P causes S" is supported by counterfactual conditional statements.
(2) Symbol S (instances of symbol S) conveys information about property P if "P causes S" is a law.

In Vlad's case, symbol S is the mental representation of vodka, which expresses the property of being vodka, provided that the generalization "the property of being vodka causes the appearance of the VODKA representation" is supported by counterfactual conditional statements. An example of such a counterfactual conditional statement is, "If a bottle of vodka were to appear now in Vlad's presence, it would elicit his representation of vodka." The second condition states that Vlad's representation of vodka conveys information about the property of being vodka, provided that the statement: "The property of being vodka causes the appearance of the representation of vodka," is a law (or a generalization justifying counterfactual conditional statements), meaning that the presence of vodka in Vlad's presence causes the appearance of the corresponding representation of vodka in Vlad.

However, it is not known how to distinguish type 1 situations from type 2 situations. The normal conditions thus defined are quite difficult to apply in practice. Fodor notes that even an extreme idealization, according to which systems in type 1 situations are infallible, would not help in resolving the disjunction problem.

> God, by definition, doesn't make mistakes; His situation is always Normal. But even God has the disjunction problem on the assumption that the content of His thoughts is determined by their causes and that some of His thoughts are caused by some of His others. The sad moral is, we still have the disjunction problem even after we idealize to infallibility. [11: 81]

Distinguishing between these two types of situations is therefore a dead end. Moreover, the causal or correlational relationship itself does not prevent a situation where the extension of a symbol expands, as I have shown using the example of drunken Vlad. Consequently, error becomes impossible, since the representation is always correct, and its reference takes the form of an awkward disjunction. Normal conditions, therefore, do not protect the informational-causal conception from the disjunction problem.

5.1.6 Solutions to the Problem of Reference Determination. Teacher or Expert

Normal conditions can also be explicated through the knowledge of an infallible teacher [2] or expert [11].[6] A teacher-coach can teach indicating the correct referent, but also reward for associating the symbol with the correct referent and punish for indicating the incorrect referent. Shifting the burden of proof onto expert knowledge assumes that we know what the experts themselves refer to, but does not explain how this happens. Logically, it is possible to imagine an oracle-expert who does not represent, but merely gives a sign of 'yes' or 'no'—depending on what they consider the correct answer, and what not (in a radical version, uses a stick as a sign of a false answer). Nevertheless, appealing to expert knowledge assumes intentionality but does not explain it. Intentionality is not explained in terms other than itself, since the ultimate point of reference for the student is to be the expert's knowledge, however minimal, expressed by means of an appropriate sign.[7]

The assumption of expert knowledge seemingly resolves the dilemma. However, a new problem emerges, known as symbolic theft [18]. This issue originates from Artificial Intelligence and concerns the creation of new categories in artificial systems. These categories refer to symbols that have already been grounded by another system. Thus, the symbol grounding problem remains unsolved if the origins of these grounded symbols are unknown. Similarly, when the knowledge of a teacher or expert determines reference, the problem of determining reference is not resolved. It would first be necessary to explain the source of the expert's knowledge, which would require the naturalization of intentionality. Otherwise, the dilemma remains unresolved.

In summary, the concept of normal conditions is constructed in such a way that it presupposes what should be demonstrated, namely which causal relations are significant for being the referent of a given representation. Establishing reference is not possible within a conception that relies solely on causality or correlation. These relations do not distinguish representationally significant relations. Moreover, referring to the knowledge of a teacher or expert assumes an understanding of the intentions and the expert's knowledge about referents. Consequently, one cannot explicate intentionality in terms other than itself. This leads to a dilemma: either intentionality is not explicated, or it is presupposed. In both cases, intentionality is not naturalized.

Solutions positing type-1 situations, where the logical necessity of infallibility is assumed, seem akin to foundationalism, which posits an indubitable foundation of knowledge (in this case, an unquestionably certain basis for the correctness of representation). A known opponent of foundationalism, defended by members of the Vienna Circle such as Moritz Schlick, was Otto Neurath, who likened scientific

[6] Normal conditions can also be explicated through the concept of function [11]. I discuss the concept of function more in the chapters dedicated to teleosemantics.

[7] In a conception enriched by an additional social factor, this leads to problems associated with the division of linguistic labor [6].

knowledge to a boat being rebuilt at sea [19: 92]. This vision of knowledge, devoid of a uniform foundation, was later defended by Willard Van Orman Quine [20]. Wilfrid Sellars also conducted a critique of this position [21]. The views of contemporary naturalists, which could be encapsulated by Sellars' formula "science is the measure of all things," exclude the foundationalist stance, as no current scientific theory constitutes an indubitable foundation of all knowledge, and unceasing criticism is a characteristic feature of scientific thinking. Therefore, any current theory may undergo significant modification or be rejected in the future. For this reason, foundationalism is usually considered incompatible with naturalism, which inherently challenges anything that would constitute such a foundation. A weaker form of foundationalism, which acknowledges the fallibility of fundamental judgments, might be acceptable to naturalism, but unfortunately, the feature of infallibility is characteristic of type-1 situations. There, for conceptual reasons, there can be no fallibility. Thus, if informational-causal semantics is to be naturalistic, it must be able to reject foundationalism.

All these difficulties in describing the possibility of misrepresentation stem from the desire to maintain the concept of misrepresentation while defining intentionality in terms of a binary relation—causal or correlational. The absence of one of the elements means that this relation does not occur. Therefore, if mental representation, characterized causally or correlationally, refers to a non-existent referent, it is by definition not a representation. This is precisely why error is excluded by a purely extensional conception and, in fact, reduced to an absence of representation. In other words: distinguishing falsehood from non-representation is impossible on the grounds of such a conception, which is based solely on causal or informational relations. By the very nature of these relations, error is indistinguishable from the absence of representation.

5.1.7 Conceptual Application on Examples

I will illustrate the problem of determination of reference with selected examples from mind and brain sciences. I will bypass the issue of trivialization, as it concerns the danger that everything is representational, and the examples discussed here are already associated with phenomena recognized in science as uncontroversially representational. I discuss the disjunction problem and the hard problem of content, which lead to troubles with misrepresentation. These problems typically arise in the context of learning, and more broadly in acting on representations, whether they are structural representations (cognitive map, mental imagery) or symbolic (the concept of rombix).

Rombix

In the informational-causal theory, the reference of the concept ROMBIX is identified with its content. In Lewicki's study, the experimenter classifies various cards as rombices or non-rombices, and the subject learns to recognize them based on the experimenter's reactions, who affirms the subject's correct response or negates the subject's incorrect response.

Imagine that we are observing the experiment at the moment when the experimenter places one of the cards in a row, and based on it, the subject must decide whether it contains a rombix or not. A given card may depict a figure of rombix, causing the symbol ROMBIX to appear in the subject's mind, or not.

Causality or correlation alone do not provide a basis for distinguishing referents that fall under the concept ROMBIX from other figures that do not. Therefore, the subject encounters the disjunction problem. They could describe every figure as "rombix or non-rombix," and since that is a tautology, they could never be mistaken. At the same time, it is difficult to describe how it is possible for the subject to learn to classify a given concept, learning from mistakes, when there is no criterion of distinction for misrepresentation from the absence of representation.

An advocate of the informational-causal theory might respond that the subject could benefit from the experimenter's knowledge, who provides the correct answer every time the subject makes a mistake. However, this does not solve the disjunction problem, as it assumes that the experimenter somehow knows what the referent of the concept ROMBIX is. This would be a version of an unsuccessful attempt to solve the disjunction problem through expert knowledge.

One could argue, however, that utilizing the experimenter's knowledge is justified. This is, after all, the assumption of the classical theory of concepts, which acknowledges that we learn concepts from others. However, this is not a satisfactory solution in the theory of reference, which aims to reduce the intentional to the non-intentional. While we acquire knowledge from others, a regress of learning cannot be endless, as it would require an infinite chain of authorities. There must be a way to indicate how the first teacher has a determined reference for their concept ROMBIX. The information-causal account, which is conceptually impoverished, fails to address this task satisfactorily and, as a result, does not naturalize intentionality.

Cognitive Map

In Tolman's experiment, a rat is trained to find food at the end of a maze. When transferred to a different maze without food, it appears that the rat is surprised by the absence of food. A scientist can conclude this not only from the rodent's behavior but also from signs of dopamine secretion in the animal's brain.

In terms of the information-causal theory, one could say that the rat behaves differently in the second maze. Proponents of the Wisconsin-style semantics may refer to the relationship with the environment (in this case, with the maze). Intuitively, we would say that the rat realizes it is not in the first maze and therefore has made a mistake. However, within the Wisconsin concept, such a description is difficult to justify because it struggles with the disjunction problem and, above all, does not allow for the discussion of error, as error is equated with non-representation. If so, in the second maze, the rat does not misrepresent the first maze as much as it lacks a representation of the first maze, or at best, has a representation of "maze 1 or maze 2." For this reason, this conception leads to even less intuitive consequences than internalism; although internalists exclude the relationship with the environment, they can still assert that the rat in the second maze would have the same representation. Therefore, externalism in the Wisconsin version is not significantly better than the internalist conception.

Mental Imagery

Let's set aside the difficulties with determining reference. It is worth noting that the causal conception does not require that referents currently exist; as long as they existed in the past, they can be the referent of a representation. Thus, Vlad may still have a representation of his dachshund, which died five years ago.

The subject in the mental rotation experiment finds themselves in a similar situation (see Fig. 5.1). The visual stimulus generates a certain representation, which they must manipulate to verify whether it indeed represents the figure from the second image. Therefore, it is not crucial that the figure, as mentally rotated, does not currently exist before the subject's eyes, since the subject has already seen it once.

An analysis of selected examples shows that Wisconsin-style semantics, leading to disjunction problems and the hard problem of content, does not adequately describe the manipulation of mental imagery, the classification of concepts, or learning from mistakes.

If it were not for the problems with determining reference, the informational-causal Wisconsin-style theory would not fare so poorly. In principle, it accounts for representations of currently non-existent objects. However, due to the indeterminacy of reference, it becomes entangled in profound difficulties.

5.2 Fodor's Asymmetric Dependency Account

Fodor's account represents the most sophisticated variant of the informational-causal conception and an interesting attempt at accounting for the possibility of error. Although Fodor relies on causal relations, he uses asymmetric dependency, the fundamental element of the new conception, precisely to describe misrepresentations.

I will briefly discuss the main features of Fodor's asymmetric dependency theory (Sect. 5.2.1). Then, I will present its strengths and weaknesses (Sect. 5.2.2), examining some consequences of this approach.

5.2.1 Introduction to Fodor's Asymmetric Dependency Account

Motivations

In the 1980s, Jerry Fodor radically changed his views on the proper concept of meaning in psychology [3, 11, 22]. He abandoned the methodological solipsism he had previously defended, which included the thesis that psychology could not address the reference of representations [23] (see also Chap. 3 of this book). On the contrary, from the mid-1980s, he began to develop an informational-causal conception of reference.

Like proponents of informational-causal semantics, Fodor understands meaning in naturalistic terms. The referent is linked to the representation by probabilistic relations. What distinguishes his approach from Wisconsin-style semantics is the condition of asymmetric dependency.

Fodor summarizes his conception of meaning in three conditions. It is worth noting that conditions 1–3 are only sufficient for the naturalization of intentionality, not necessary. In other words: Fodor does not rule out that it can be naturalized differently. He claims that a representation of X refers to X, where Xs are the referred to properties, and Ys are not, if:

1. '*X*s cause "*X*"s' is a law.
2. Some "*X*"s are actually caused by *X*s.
3. For all *Y* not = *X*, if *Y*s qua *Y*s actually cause "*X*"s, then *Y*s *causing* "*X*"s is asymmetrically dependent on *X*s *causing* "*X*"s [11: 121].

One might find the formulation "Ys, qua Ys, cause Xs" somewhat puzzling. Fodor assumes that the causal relation occurs by virtue of natural law between properties. Thus, if table salt, by having the property of being sodium chloride, causes the representation WHITE POWDER, then such a causal relation, according to Fodor, cannot constitute the meaning of the representation WHITE POWDER. This is because there are counterfactual conditionals regarding the causation of the representation WHITE POWDER by sodium chloride that remain true even when a given portion of table salt is not sodium chloride (e.g., on Twin Earth). In other words: Fodor requires that the referent of the representation was that which, by virtue of all its essential properties, elicits that representation. It is undeniable that his conception has not only strong assumptions about the nature of causality (as occurring between properties) but also strong ontological assumptions, as Fodor presupposes the existence of essential properties and universals. Not all contemporary conceptions of causality share these assumptions. However, I will assume that this aspect of Fodor's view is relatively uncontroversial, as its deeper flaws lie elsewhere.

The first condition states that representations of Xs are caused by Xs. In informational-causal semantics, condition 1 is equivalent to the claim that representations of Xs carry information about Xs; this is asserted, for example, by Dretske [2]. It is worth noting that representations of Xs do not carry information exclusively about Xs. The first condition does not preclude their tokening for other reasons.

Fodor frames laws in terms of counterfactual dependencies ceteris paribus, which allow for exceptions under circumstances other than normal. If organisms with the concept of SALT were to emerge, then under additional conditions (e.g., unimpaired perception), the presence of salt would elicit the tokening of the representation SALT in them. Nevertheless, the counterfactual dependency itself does not guarantee that the representation SALT exists or will always arise in the possessors of this representation in the presence of salt.

Laws, understood as ceteris paribus (all other things being equal) relations, despite allowing exceptions, should be formulated to be as general as possible. Thus, it would not be a law that the table salt from my salt shaker dissolves in water, but rather that every portion of table salt dissolves in water. Defining representations solely through the nomic dependencies of the representation "table salt" and salt would result in too many generalizations becoming essential for the referent. Moreover, it does not allow solving the disjunction problem, i.e., the tokening of representations, such as "table salt," in the presence of an object or its features that do not belong to the reference of the given representation, such as sugar.

Condition 2 is intended to narrow the overly broad scope of generalizations that determine the reference of representations. It takes into account the actual history of tokens of X, thereby excluding strange examples from Twin Earth, such as water being XYZ, because they are inconsistent with the actual history of its tokens.

Condition 3, which is crucial for the concept of error, involves certain difficulties. It speaks of an asymmetric dependency, according to which the tokening of the representation "table salt" in the presence of sugar depends on the tokenings of "table salt" in the presence of table salt, but not vice versa. Similarly, the tokening of a representation in the face of an image of an object (e.g., an image of a king) is asymmetrically dependent on the tokening of the representation of the object in its presence (e.g., a king). This dependency holds true when, for instance, some instances of sugar indeed cause the appearance of the representation "table salt," as well as under the assumption of the truth of counterfactual conditional statements, according to which sugar would not usually cause the occurrence of "table salt," but every portion of table salt would cause the representation of salt. The asymmetric dependency not only accounts for intuitions about the concept of representation but also serves as a solution to the disjunction problem. Below, I will analyze the advantages and disadvantages of this solution.

The asymmetric conception assumes that at least sometimes a sign refers correctly, which allows us to rule out a complete mistake at all times. In addition to correct occurrences of the sign in the history of use, it admits a partial mistake, even at the first use of a representation: even if it turned out that 'table salt' sometimes appears as a result of contact with a substituted substance from Mars, it does not mean that sometimes "table salt" refers to table salt or the substance from Mars.

5.2.2 *Disadvantages and Advantages of the Asymmetric Conception*

Fodor's asymmetric theory circumvents some of the problems inherent in purely extensional semantics. As such, it has an advantage over causal-informational theories, which rely on normal conditions. Moreover, Fodor's asymmetric theory fulfills the crucial desideratum of the possibility of representational error. Condition 2 does not exclude error; it avoids problems with the assumption, along the lines of functionalism, of infallibility under normal conditions, that is, in type 1 situations. As I have emphasized earlier, Fodor explicitly criticized the distinction between these two types of situations in causal theories of reference. In his view, it is entirely unnecessary. Since the law is ceteris paribus, and it is only required that at least some representations be accurate, there is no problem with error even under normal conditions. Condition 3 elegantly captures the dependency of erroneous representations on correct ones. Thanks to this, Fodor's theory is decidedly closer to the naturalization of intentionality than the Wisconsin-style semantics.

Furthermore, Fodor's later conception applies to representations concerning phenomena of all kinds, not just linguistic ones. This is because Fodor rejects his earlier position that all mental representations are linguistic in nature. As a result, his new conception meets the desideratum of universality (I.2). This is achieved by the picture principle.

The picture principle states that pictorial representations, like linguistic ones, have a structure unique to themselves. It is formulated as follows:

> Picture principle: If P is a picture of X, then parts of P are pictures of parts of X. [24: 173][8]

Although Fodor acknowledges structural properties as important features of mental representation, he does not elaborate on the dependency of reference on structure. Therefore, it is unclear why some structural properties are significant for reference while others are not. Meanwhile, it is precisely the relationship between structural properties and reference, which allows us to determine which structural properties define content, that is essential to the nature of the group of representations known as structural representations [16, 25]. Structural properties not only indicate properties of the environment but also allow for their description, thereby becoming meaningful for the functioning cognitive system.

The picture principle pertains to the structural dependencies of structural representations on the properties of the environment, thereby allowing for a correspondence of representations with the environment. However, in the absence of semantic rules that bind structural representations to environmental properties, it is impossible to determine how such correspondence arises. Consequently, it is not feasible to ascertain which features of representations of a given environment are its significant features. Thus, the picture principle fails to solve the hard problem of content—it does not explain why certain correspondences are significant for representation. In this respect, Fodor's conception is no different from Wisconsin-style semantics.

The asymmetric conception also leads to other problems with the concept of representation. For instance, Fodor assumes that basic representations refer by virtue of causal relationships, and therefore their referents must exist. According to the assumption of causal conceptions—which is shared with classical empiricism—empty representations arise solely from the composition of representations

[8] The compositional principle for images differs from the compositional principle for sentences. While all parts of an image are also its constituents, not all parts of a sentence are its constituents. For instance, in the sentence: "Even though John has nice teeth, he must visit the dentist," the word "even" is not part of the sentence, but the words "even though" are.

that already have referents, precluding the possibility of basic (non-decomposable) empty representations. It is thus claimed that fictional representations are composites of representations of properties that exist in the world, for example, a SQUARE CIRCLE is composed of the representations of a square and a circle. However, one might question whether all such representations are composites: if Fodor considers concepts like BOOK and CARBURETOR to be basic and non-decomposable, why would concepts like DEMON, ECTOPLASM, and CHAKRA be decomposable?

Since Fodor additionally adopts an ontology of properties, in his semantics all terms are abstract (in this regard, "cow" and "number" are no different), which spares him from entangling in explanations of how concrete representations become abstract representations. This is a strong assumption, implying that even proponents of realism would have to accept that all concepts are abstract. The conception of mental representations should be metaphysically neutral. Yet, Fodor, in his conception of representations, arbitrarily settles a metaphysical dispute without providing any justification.

I will now address another criticism of the conception. Bickhard accuses Fodor of proposing an asymmetric conception that is too permissive. In this way, he critiques his approach to representational error. Bickhard [29] provides the example of a virus particle that mimics the action of a neurotransmitter. In both cases—during infection and with the genuine neurotransmitter—there is a causal and nomological dependence between the neuron's response and the receptor's action, and an asymmetric dependence between the virus and the neurotransmitter is also fulfilled. However, this constitutes only a functional error, which is not a cognitive error (the virus represents nothing), as one would expect in a theory of representation. As Bickhard points out, this does not allow us to grasp the nature of representational dependencies.

The problems with Fodor's conception are also well illustrated by all sorts of functioning forgeries. For instance, a lock pick that counterfeits a key to a lock is asymmetrically dependent on the proper key, just as Reebok knockoff shoes are asymmetrically dependent on the original, but this is not a dependence of an epistemic nature, rather it is functional.

A proponent of Fodor might respond to this criticism by asserting that viruses (from Bickhard's example), keys, or shoes are not representations. It is clear that they are not mental representations, and thus should be excluded by conditions 1–3. However, these conditions do not exclude from the category of representations anything except objects or features that are not the result of any cause or occur randomly (irregularly, and thus not by law). Meanwhile, keys regularly open doors, just as lock picks do; expensive brand shoes and their counterfeits are regularly the results of known manufacturing processes. Fodor's theory does not disqualify them a priori, allowing for their representational nature. Many proponents of causal theories emphasize, after all, that all causal interactions convey information (as Dretske does, among others)—the placement of a shoe on the floor indicates where it is located. In this sense—quite minimal—it is a representation. As long as Fodor lacks a better criterion for being a representation, he cannot dismiss these examples. The criticism thus remains valid, and Fodor's conception cannot defend itself against the more general charge: that it trivializes the concept of representation by defining it too broadly. His approach, therefore, does not address the problem of trivialization, just as Wisconsin-style semantics does not.

An undeniable advantage of Fodor's approach is the inclusion of an error detectable by the organism [11]. In none of his later works does Fodor develop his assertion that organisms can only have representations if they are able to make their corrections upon detecting an error. Specifically, he does not show how this relates to the concept of representation based on asymmetric dependency.

5.2.3 Application of the Concept to Examples from Mind and Brain Sciences

The problem of reference determination in the asymmetric version of Fodor's theory will be illustrated with selected examples from mind and brain sciences. I will address the disjunction problem and the hard problem of content.

Rombix

In Fodor's account, the content of the concept ROMBIX—as in information-causal views—is its reference. Fodor might argue that there is a probable dependency between the learner who is learning the concept ROMBIX and rombices, distinguishing referents that fall under the concept ROMBIX from other objects. The answer to the question of what kind of dependency this is, according to Fodor, is provided by empirical evidence, not philosophy. In the case of a made-up concept such as ROMBIX, the teacher knows its denotation and is therefore a source of knowledge for the learner. However, this solution does not have the undesirable consequences that arise from adopting normal conditions by Wisconsin-style semantics. The asymmetric dependency of incorrect representations on correct ones, expressed by condition 3, does not exclude the possibility that the learner may make a mistake while learning the concept.

Since misrepresentations of rombix are asymmetrically dependent on correct representations, knowing the features of rombix is sufficient to recognize something as a misrepresentation of rombix. It seems that Fodor thus deals with the (vertical) disjunction problem, as correct representations are privileged over incorrect ones.

Mental Imagery

In their early works on mental representations, Fodor and Pylyshyn recognized only linguistic representations; they treated mental imagery in the same way [26, 27]. In the asymmetric conception, Fodor allows for structural representations, including mental imagery, acknowledging their non-linguistic character.

According to the picture principle, mental images are compositional: if an image represents an object, such as the projection of a polyhedron onto a plane, then the parts of this image, i.e., the sides of the imagined figure, correspond with the corresponding parts of the depicted object, i.e., the sides of the figure in the drawing.

The properties of mental imagery (e.g., a mentally rotated polyhedron) asymmetrically depend on the properties of an actually existing object (as depicted in the illustration). The reference of the representation of the imagined figure depends on the properties of the actually existing figure. Therefore, asymmetrical dependence allows Fodor to treat mental imagery as representational.

Cognitive Maps

Fodor's asymmetrical concept also fares better with another example of structural representations, namely cognitive maps. Thanks to the picture principle, it can be said that cognitive maps reflect the structure of the environment. For instance, a rat, through place cells, can reflect the structure of the maze it is in. Place cells reflect the properties of the maze through a probabilistic causal relationship. Since Fodor's approach (like any information-causal theory) includes reference, the rat's representations relate to the environment depending on where the rat is located. Thus, it can be said that the rat behaves differently in another maze.

Fodor's causal semantics, like Wisconsin-style semantics, is a conception of reference, but not intentionality. Therefore, although it is possible to say that the rat reflects the environment depending on where it is located, it is not possible to determine which properties of the environment are reflected by it. It can be stated that the rat represents the maze to which it has been transferred, but it cannot be explained which properties of the representation are essential for the rat to represent that particular maze. The picture principle does not help—while it allows for pointing out the importance of the structure of the representation, it is unclear where the correspondences come from. Thus, the asymmetrical concept, like Wisconsin-style semantics, cannot solve the hard problem of content.

A misrepresentation of a rat, which erroneously represents a foodless maze as one containing food, may occur because there is a probabilistic causal relationship linking the type of environmental representation (map) with an environment of given properties. The maze in which the rat does not find food causes a misrepresentation, which yet is asymmetrically dependent on the accurate representation. Within the framework of the asymmetrical theory of dependence, it is not known how it happens that the rat actually made a mistake and realized that it did. Nevertheless, the asymmetrical dependence of erroneous representations on accurate ones allows for the inclusion of misrepresentation.

Fodor provides a criterion for distinguishing between misrepresentation and the absence of representation, thereby addressing the vertical disjunction problem and allowing for a satisfactory account of the manipulation of mental imagery. However, similar to Wisconsin-style semantics, the asymmetric conception cannot meet the hard problem of content, as it does not explain the origin of satisfaction conditions. Since Fodor's conception does not explain the emergence of satisfaction conditions, it cannot elucidate how an organism can be capable of representational error.

5.3 Summary

In the first part of the chapter, I analyzed the informational-causal semantics in the Wisconsin style, and in the second part, Fodor's asymmetric conception as a certain version of this semantics. The informational-causal semantics does not solve the problem of determining reference or the disjunction problem. Fodor proposes a better solution to these difficulties, but it is not sufficient to defend the informational-causal account of reference against the accusation of trivializing the concept of representation.

The informational-causal theory does not fulfill the desideratum of determining reference and content (II.2.1), nor does it meet either of the two desiderata concerning representational error (II.3.2 and II.3.3). Fodor's conception, on the other hand, takes into account the possibility of representational error (II.3.2), although his account of representational error raises some doubts—it is purely functional rather than cognitive. Therefore, none of the causal theories discussed in the chapter fulfills the desideratum II.2 about the naturalization of intentionality or II.3 about an adequate theory of content. Cummins and Ramsey argue for the colossal significance of structural representations; according to Fodor, however, their structures have no significance in the concept of reference. It is questionable whether the conception of reference alone—without a connection to structure—is sufficient to naturalize intentionality. In the next chapter, I examine whether the externalist conception, which takes into account not only reference but also truth conditions (overlooked in the informational-causal account), is capable of naturalizing intentionality and providing an adequate theory of content.

References

1. Sterelny, K.: The representational theory of mind: an introduction. Blackwell, Oxford, Cambridge (1990)
2. Dretske, F.I.: Knowledge and the Flow of Information. Cambridge University Press (1981)
3. Fodor, J.A.: Semantics, Wisconsin style. Synthese **59**(3), 231–250 (1984). https://doi.org/10.1007/BF00869335
4. Stampe, D.: Toward a causal theory of linguistic representation. Midwest Stud. Philos. **2**(1), 42–63 (1977). https://doi.org/10.1111/j.1475-4975.1977.tb00027.x
5. Kripke, S.: Naming and Necessity. Harvard University Press, Cambridge, Mass (1980)
6. Putnam, H.: The meaning of meaning. In: Philosophical Papers: Mind, Language, and Reality, vol. II, pp. 215–271. Cambridge University Press, Cambridge (1975)
7. Lettvin, J.Y., Humberto, H.R., Maturana, R., McCulloch, W.S., Pitts, W.H.: What the frog's eye tells the frog's brain. Proc. IRE **47**(11), 1940–1951 (1959)
8. Schulte, P.: How frogs see the world: putting Millikan's teleosemantics to the test. Philosophia **40**(3), 483–496 (2012). https://doi.org/10.1007/s11406-011-9358-x
9. Quine, W.V.O.: Word and Object. MIT Press, Cambridge (1960)
10. Fodor, J.A.: Why paramecia don't have mental representations. Midwest Stud. Philos. **10**(1), 3–23 (1986). https://doi.org/10.1111/j.1475-4975.1987.tb00532.x
11. Fodor, J.A.: A Theory of Content and Other Essays. MIT Press, Cambridge (1992)
12. Godfrey-Smith, P.: Misinformation. Can. J. Philos. **19**(4), 533–550 (1989)

13. Mackie, J.L.: The Cement of the Universe: a Study of Causation. Clarendon Press, Oxford (1974)
14. Woodward, J.: Making Things Happen. Oxford University Press, Oxford (2003)
15. Frege, G.: Über Sinn und Bedeutung. Zeitschrift für Philosophie und philosophische Kritik **100**, 25–50 (1892)
16. Ramsey, W.M.: Representation Reconsidered. Cambridge University Press, Cambridge (2007). https://doi.org/10.1017/CBO9780511597954
17. Harman, G.: Thought. Princeton University Press, Princeton, NJ (1973)
18. Cangelosi, A., Harnad, S.: The adaptive advantage of symbolic theft over sensorimotor toil: Grounding language in perceptual categories. Evol. Commun. **4**(1), 117–142 (2002). John Benjamins Publishing Company
19. Neurath, O.: Protocol statements. In: Cohen, R.S., Neurath, M. (eds.) Philosophical Papers 1913–1946, pp. 91–99. D. Reidel Publishing Company, Dordrecht (1983)
20. Quine, W.V.O.: Two dogmas of empiricism. Philos. Rev. **60**(1), 20–43 (1951). https://doi.org/10.2307/2181906
21. Sellars, W.: Empiricism and the Philosophy of Mind. Bobbs-Merrill, Indianapolis (1956)
22. Fodor, J.A.: Psychosemantics: the Problem of Meaning in the Philosophy of Mind. MIT Press, Cambridge (1987)
23. Fodor, J.A.: Methodological solipsism considered as a research strategy in cognitive psychology. Behav. Brain Sci.. Brain Sci. **3**(1), 63–109 (1980). https://doi.org/10.1017/S0140525X00001771
24. Fodor, J.A.: LOT 2: The Language of Thought Revisited. Clarendon Press, Oxford, New York (2008)
25. Cummins, R., Roth, M.: Meaning and content in cognitive science. In: Schantz, R. (ed.) Prospects for Meaning, pp. 365–382. de Gruyter, Berlin, New York (2012)
26. Fodor, J.A.: The Language of Thought. Thomas Y. Crowell Company, New York (1975)
27. Pylyshyn, Z.W.: The imagery debate: analogue media versus tacit knowledge. Psychol. Rev. **88**(1), 16–45 (1981)
28. Millikan, R.G.: Biosemantics, The Journal of Philosophy 86(6), 281–297 (1989)
29. Bickhard, M. H.: The interactivist model. Synthese, 166(3), 547–591 (2009)

Chapter 6
Truth-Conditional Semantics

In this chapter, I analyze extensional conceptions of meaning as truth conditions. Truth conditions apply to descriptive representations, which can be true or false. Some descriptive representations are judged solely on veridicality, as in the case of perceptual representations. Meanwhile, descriptive-directive representations can be realized or unrealized. Each of these types of representations can be satisfied or not, if the representation is accordingly: true, accurate, or realized.[1] Truth conditions are a paradigmatic example of satisfaction conditions.

Conceptions of content as satisfaction conditions can be seen as another variant of externalism, one that does not appeal to the concept of reference (but does not exclude it). In other words: wide content can be treated not only as reference (as in informational-causal conceptions), but also as satisfaction conditions for mental representation. It is worth considering whether the mere framing of intentionality in terms of satisfaction conditions suffices for a proper understanding of content and whether it resolves Hutto and Myin's hard problem of content, which requires that mental representations have satisfaction conditions not by virtue of a relationship with other mental representations.

In Part 1 of the chapter, I analyze the consequences of conceiving meaning (content) as entirely reducible to satisfaction conditions. For the purposes of the chapter, I call this approach a "purely extensional conception of meaning as truth conditions." In Part 2, I invoke Davidson's conception of meaning. In this conception, the satisfaction conditions alone do not determine the meaning, but are to be linked in a unified theory of meaning and action [1]. The relationship between truth and meaning is described in terms of triangulation. Davidson also recognizes representational error as essential for representation.

Davidson does not carry out the naturalization of mental content, rejecting the possibility of reducing the content of linguistic phenomena to non-linguistic ones,

---

[1] Ruth G. Millikan characterizes the content of descriptive-directive representations in terms of conditions of satisfaction, which are crucial for her teleosemantics. More on this topic can be found in Chap 8.

K. Bielecka, *What is Misrepresentation?*, Studies in Applied Philosophy, Epistemology and Rational Ethics 71, https://doi.org/10.1007/978-3-031-84375-4_6

characteristic of humans and animals. Adopting a naturalistic perspective, I add to Davidson's conception of meaning the assumption that such a reduction is possible. Therefore, I propose to use the term "Naturalized Truth-Conditional Semantics" (NTCS) for this modified version of Davidson's conception.

In this chapter, I continue the search for an adequate conception of content. I analyze the truth-conditional semantics primarily due to the desideratum of naturalizing intentionality (II.2) and the desiderata concerning errors (II.3.2, II.3.3). Truth conditions are a paradigmatic example of satisfaction conditions. Therefore, prima facie, the semantics of truth conditions should fulfill the desideratum:

II.2.2. A satisfactory theory of representation demonstrates why a representation can bear content.

However, I show that the truth-conditional semantics does not fulfill this desideratum in a way that addresses the hard problem of content. The conclusion of this chapter is that another conception must solve the hard problem of content in order to speak of the naturalization of intentionality.

6.1 Purely Extensional Conception of Meaning as Truth Conditions

In Sect. 6.1.1, I present a purely extensional conception of meaning as truth conditions.[2] In Sect. 6.1.2, I analyze the objections of Donald Davidson, Gilbert Harman, and John Foster, and consequently reject this conception. Although the objections are serious, I demonstrate through examples from scientific practice that this conception may have certain advantages; among others, its formal and precise nature facilitates the differentiation of various types of representation. Thus, while this conception does not naturalize intentionality, it may play a significant role in scientific research and philosophical discussions (Sect. 6.1.3).

At the same time, I show that the truth-conditional conception (also in the NTCS version, see part 6.2) is applicable to the description of non-linguistic structural representations and memory processes, as illustrated by cognitive maps in rats.

6.1.1 Introduction to the Conception. Motivations

The goal of proponents of construing meaning in terms of truth conditions is to create a universal conception of meaning in natural language using (solely or among

[2] For simplicity, I only deal with truth conditions, not conditions of satisfaction. If a conception cannot account for the content of descriptive representations, it cannot account for the content of descriptive-directive representations.

other things) the formal tools of extensional logic [2].[3] However, I transform it into a general conception of mental representations.

The fundamental idea of construing meaning as truth conditions can be captured in the slogan that "knowing the meaning of a sentence is to know the conditions under which the sentence is true." Knowing the meaning of a sentence (e.g., "The grass is green"), we can determine when it is true (when the grass is indeed green). I propose to develop this scheme into two versions: strong (P1) and weak (P2). Let us assume that s stands for a sentence, and p for a truth condition. In its strong version, the scheme of knowing meaning as truth conditions is as follows:

(P1) s means that p if and only if s is true if and only if p.

In its weaker form, it is expressed as follows:

(P2) s means that p only if s is true if and only if p.

In its strong form, as articulated by schema P1, we treat the main thesis as an equivalence, and thus understanding of meaning is fully captured by the knowledge of the sentence's truth conditions. In the weaker version, however, the thesis expresses an approximate equivalence, meaning that knowledge of truth conditions is necessary but not sufficient for understanding meaning. The weak version allows for an interpretation according to which truth conditions are supplementary to the comprehensive understanding of the sentence's meaning, while meaning is not fully accounted for by truth conditions.

Discussing the conception of meaning as truth conditions, I examine the thesis in both its forms. According to the strong thesis, meaning is entirely given by truth conditions. The strong form of the thesis does not address the reference of representations or their components, which—as I have argued in previous chapters—is an essential aspect of meaning. Therefore, I find the thesis in its weaker form more convincing.

The theory of meaning, according to the strong thesis, should be finite, axiomatizable, and recursive, similar to Tarski's theory, with the difference that truth is not defined but taken as primitive. The principle of compositionality and recursion are fundamental principles for creating meanings of complex expressions from the meanings of individual words; they play a significant role in this conception. The principle of compositionality allows us to understand how the truth of a sentence depends on the meanings of its constituent parts (names, predicates) and the way they are combined through rules for the assignment of constituent parts to other constituent parts (including the truth table). Having syntactic rules for creation is the first step towards understanding the meanings of sentences in a given language L [1]. Thus, the discussed concept meets the desideratum of productivity. It allows for the

[3] The goals of the conception of meaning as truth conditions are the opposite of the goals of Tarski's theory, where truth is explained, and the concept of meaning is taken as given [3]. I am inspired by Davidson's project in those works where he has not yet attempted to create a unified theory of meaning and action [2, 4].

use of logical means (the aforementioned recursion), ensuring compositionality and systematicity, which is considered a credible explanation for productivity.

The next step is to create the semantics of language L: having derived the meanings of the language's primary expressions (dictionary) based on their occurrence in sentences and knowing the interpretation of the formal rules of language L (rules for the composition of meanings), the theory generates interpretations for sentences, for example:

(T) 'The grass is green' is true if and only if the grass is green.

The right side of a T-sentence defines the truth conditions of the sentence and the meaning of the sentence on the left side.

6.1.2 Objections to a Purely Extensional Conception of Truth Conditions

In this section, I discuss three interrelated objections: the slingshot argument and the associated objections of Foster and Harman, which reveal weaknesses in the described conception. The consequences of the slingshot argument stem from its purely extensional nature. I also show that the slingshot argument is a manifestation of the difficulty in unambiguously determining meaning.

In informational-causal conceptions, meaning cannot be captured because one sentence or symbol corresponds to many referents, which is the essence of the disjunction problem; in the semantics of truth conditions, meaning cannot be captured because many sentences correspond to one meaning (slingshot). Ambiguities of this kind testify to the impossibility of capturing meaning in a purely extensional way.

Slingshot Argument

The slingshot argument[4] was formulated by Davidson [4, 6]. It is aimed against purely formal extensional conceptions, whose weaknesses it is supposed to demonstrate.

The slingshot argument targets two formal assumptions of conceptions that reduce meaning to logical value: (1) the principle of substitutivity, which is based on the idea that the logical value of a sentence remains unchanged after substituting any word or expression in the sentence with any word or expression of the same reference; and (2) the assumption of referential identity of logically equivalent sentences (Frege's assumption that true sentences refer to Truth, and false sentences to Falsity). By these assumptions, meaning is identified with reference. All true sentences have a common reference [4].

Consider two true sentences: sentence A ("Snow is white") and sentence B ("Grass is green"), which correspond to facts:

(1) Snow is white

---

[4] The name of the argument originates from the work of Jon Barwise and John Perry [5].

(2) {x: x = x and snow is white} ≡ {x: x = x}
(3) {x: x = x and grass is green} ≡ {x: x = x}
(4) Grass is green

The principle of substitutivity ensures the material equivalence of sentences 1 and 4, while the assumption of logical equivalence gives the identity of the pairs of sentences 1 and 2 as well as 3 and 4.

A consequence of the slingshot argument is that all true sentences mean the same thing [1]. Indeed, two true sentences, such as "The Earth is round" and "The grass is green," cannot have different truth conditions, because their logical value is the same. Therefore, this conception does not distinguish sentences of different meaning. Thus, the true sentence:

(S) The sentence "The Earth is round" is true if and only if the grass is green.

does not express the meaning of the sentence on the right side of T-equivalence.

A purely formal extensional method of differentiating sentence meanings is too weak to narrow down the set of sentences on the right side of equivalence so that they mean the same as the sentences on the left side. The concept utilizing this method inaccurately determines the meanings of sentences, even though it does not define them in terms of logical value, but in terms of truth conditions, irreducible to logical value.

Foster's Objection

Foster's objection deepens and strengthens the slingshot argument and concerns one of the assumptions of the discussed concept. He argues that it must adopt an intensional translation relation and the interpreter's understanding of the sentence's meaning, thus presupposing what it was supposed to explain if it is to be considered a concept that aims to naturalize intentionality. In other words, this conception is circular. As Foster states:

> Obviously we cannot have it both ways: we cannot have both extensionality of syntax and the demands of translation reflected in the force of what is stated. From which it follows that either there are no truth-conditional theorems of the kind we want, or, if there are, we cannot formulate them quite as we have supposed, by the linking of structural designator and English translation. [7: 24]

As he claims:

> A T-sentence does not say that such and such a structural type would be true (meaning held constant) in all and only those circumstances in which it was the case that..., but merely that, things being as they are, this structural type is true if and only if… [7: 11]

The vacuity of the discussed conception lies in the fact that it is useless unless its user does not previously know the meaning of the sentence. As Foster points out, the user of this conception knows that the sentence "Snow is white" is true if and only if snow is white, and knows that this fact results from the theory of translation, thus it is the interpretation of the sentence on the left side of equivalence.

In summary: Foster's objection deepens and strengthens the slingshot argument. Foster defends the thesis that the only way to avoid the consequences of the slingshot argument is to refer to an intensional translation relation and the interpreter's knowledge of the meaning. However, it is still unclear how to identify T-sentences that express truth conditions and differentiate them from other true equivalences in a way that does not lead to circularity. Davidson points this out:

> For such theories [assuming detailed knowledge of beliefs or intentions accompanying individual utterances -- K.B.] use for evidence facts as subtle and hard to identify as the facts they are intended to support. [1: 5]

Harman's Objection

While Foster's objection concerning the translation relation applies to all statements, Harman's objection is specific to intensional statements, such as those about beliefs. Harman exposes a weakness in the purely extensional theory when it comes to the attribution of beliefs, a doubt similarly raised by Davidson himself.

As Harman notes, the formal tools of the discussed theory are too weak to halt the inferential process, thereby failing to disambiguate beliefs:

> (…) truth conditions are assigned to beliefs by virtue of the theory's implying clauses of the form, 'Belief *b* is true if and only if *C*'. Then the problem is that the same theory will also imply indefinitely many results of the form '*C* if and only if *D*', where '*C*' and '*D*' are not synonymous, so the theory will imply indefinitely many 'incorrect' clauses of the form 'Belief *b* is true if and only if *D*'. The problem is, in other words, that specifying truth conditions in this way does not distinguish among beliefs that are equivalent in relation to the principles of the theory. [8: 65]

The consequences of the slingshot argument for ordinary statements are such that, for example, the statement 'Snow is white' means the same as 'Grass is green' (see: the objection in point a). Harman shows that when applied to intensional statements, slightly different problems arise; namely, statements resulting from other statements will be considered mutually equivalent, even in intensional contexts. Consider the following pair of statements:

(a) I am convinced that George W. Bush voted for Barack Obama during the penultimate elections in the USA.
(b) I am convinced that the penultimate former President of the United States voted for Barack Obama during the penultimate elections in the USA.

The point is that the discussed theory allows us to equate the belief expressed in sentence (a) with the belief expressed in sentence (b), since both subordinate clauses in (a) and (b) are true, and the linguist describing my beliefs knows that the former President of the United States and George W. Bush are the same person. Belief (b) is true if and only if the subordinate clause in (a) is true, and belief (b) is true if and only if the subordinate clause in (b) is true.

Based on the concept of truth conditions, different beliefs expressed by (a) and (b) have the same content because the relation of material equivalence indicating the meaning of a belief allows for the substitution of sentences derived from (a) on the right side of T-equivalence. And the subordinate clause in example (b) follows from the subordinate clause in example (a), provided we know the additional premise that the penultimate former President of the United States is George W. Bush. If a linguist describing my beliefs knows this, they can incorporate this premise into the theory of meaning, but then such a theory of meaning may be empirically inadequate (if I do not know it). To avoid this problem, one could argue that the sentences "George Bush voted for Barack Obama during the penultimate elections in the USA" and "the penultimate former President of the United States voted for Barack Obama during the penultimate elections in the USA" are not synonymous. This is problematic for a purely extensional theory of meaning as truth conditions because, on its grounds, synonymy should be explicated in terms of truth conditions.

Furthermore, an extensional conception of meaning as truth conditions requires the naturalization of intentionality. This conception only allows us to state that meaning is somehow attributed to words or sentences, which obviously is not enough to call it an adequate concept of content. Assuming knowledge of meaning would lead to circularity. Moreover, it is unclear how to identify T-sentences expressing truth conditions and distinguish them from other true equivalences. This stems from the slingshot argument, and Foster and Harman also point this out.

6.1.3 Applying the Theory to Examples from Mind and Brain Sciences

Using examples from mind and brain sciences, I will demonstrate the application of the discussed conception. The examples illustrate its strengths and weaknesses.

Rombix

The aim of Lewicki's study was to generate a concept in the subject, hence the representation of a rombix. The experimenter shows various cards to the subject, who classifies them as rombices or non-rombices. In the discussed conception, meanings of individual words are not addressed unless they are used in sentences. However, this is not a problem because when an image is classified as a "rombix," the expression "rombix" can be paraphrased as the indicative sentence: "This is a rombix," and when the image belongs to non-rombices, "non-rombix" can be paraphrased in the sentence: "This is not a rombix."

From the premise of the experiment, it follows that the subject acquires a new concept with the assistance of the experimenter, who indicates a standard card and then tests the subject's ability to recognize the features of the concept. I suggest examining both stages of the experiment separately: (1) the concept learning stage and (2) the stage of testing the new concept in new situations (with the exposure to control cards). During the learning stage, the subject learns the concept under the experimenter's control, who knows the correct answer. The experimenter utters a declarative statement: "This is a rombix," and the subject associates his statement with the image, which can be described using the TE pattern (as I call the T-schema applied in experimental situation E), for example:

(TE1) "This is a rombix" is true if and only if a rombix is a red diamond.

Since a rombix is indeed a red diamond, the subject acknowledges the sentence: "A rombix is a red diamond." Then the experimenter shows a new card, which the subject is to classify. Suppose it is a card depicting a yellow triangle, not a rombix, hence the subject's statement: "This is a rombix," could be paraphrased as:

(TE2) "This is a rombix" is true if and only if it is a yellow triangle or a red diamond.

In the second stage of testing the concept, new images appear that the subject has not yet seen.

If the discussed objections to this theory are valid, then a subject who would act in accordance with it might believe that the meaning of the sentence: "This is a rombix," is expressed by the T-sentence:

"This is a rombix" is true if and only if the indicated image depicts a rombix.

It is worth noting that Lewicki's experiment does not exclude the possibility that some subjects may recognize only the red diamonds visible in the drawings as rombices. Indeed, learning does not have to lead to a full and unambiguous determination of the concept's meaning—as intended by psychologists. Certain variations are possible—a proponent of the truth-conditions theory as expressing the meaning of descriptive representations could thus defend it, pointing to the possibility of some indeterminacy of meaning. However, an opponent of this theory might observe that it is unclear by what means truth conditions are attributed to representations at all.

Cognitive Maps

A purely extensional conception of meaning as truth conditions aligns with the desideratum of universality (I.2). This is evidenced by the example of cognitive maps. A rat's cognitive map is a type of structural representation that is compositional, systematic, and productive [9]. As a geometric rather than logical representation, it can be described using semantics incorporating compositional mechanisms different from the standard Tarskian semantics for predicate logic (the principle of compositionality pertains to geometric relations, not logical ones). Rescorla demonstrates that there exists a formal conditional-truth semantics for cognitive maps, which does not rely on the concept of satisfaction in the Tarskian sense nor on the principle of induction [9: 399–400].

Rescorla emphasizes that thanks to the formal semantics of truth conditions for cognitive maps, which he defines in a simplified version for robots, one can show the difference between various formats of representation. This difference is based on other compositional relations. Although the semantics of truth conditions does not naturalize intentionality *per se*, it allows for the description of psychologically and neuroscientifically significant aspects of representation. Rescorla notes that this is particularly relevant to the debate over mental imagery, which can be articulated more precisely through such semantics.

Indeed, the reduction of meaning to conditions of satisfaction in a purely extensional conception is not successful, as evidenced by the objections cited, but it has certain advantages. In particular, structural representations can be distinguished from linguistic ones based on different formal descriptions of truth conditions. The structure of a representation, such as a cognitive map, is not an incidental feature but is associated with specific types of truth conditions. Unlike informational-causal theories or Fodor's asymmetric theory, where the structure of representation is unrelated to its content, in theories of truth conditions, one can capture the correspondence between the structure of representation and reality as a key feature of content. The question of the naturalization of intentionality remains open, but at least this conception allows us to differentiate various types of mental representations.

6.2 Summary

A purely extensional conception of meaning as truth conditions is limited to formal methods of describing meaning. The slingshot argument and the objections of Foster and Harman indicate that this conception cannot adequately determine content: it does not explain by what means sentences or words signify something. The formal apparatus of this conception is empty if it presupposes prior knowledge of meaning, intention, or synonymy. Therefore, this conception could not naturalize intentionality.

6.3 Meaning as Truth Conditions and Action

In Sect. 6.3, I analyze the NTCS. Davidson's framework, outlined in the text "Toward a Unified Theory of Meaning and Action" [1], is treated as the schema of the conception of mental representations, which pertains to behaviors not only of humans but also of animals (Desideratum I.2). I adapt the Davidsonian idea of action as communication, among other things, to describe memory processes. Thus, I move beyond the assumptions of Davidson's project, which did not deal with mental representations themselves, and moreover, rejected the possibility of discussing the content of animal thoughts.

According to Davidson, knowledge of truth conditions is necessary but not sufficient for understanding meaning (which corresponds to the weaker version of the schema of meaning as truth conditions P2). Therefore, he describes representations in the context of rational action in terms of triangulation (Sect. 6.3.1). Davidson's approach is the first among those discussed in this book to account for representational error detectable by the system (Sect. 6.3.2). I argue that triangulation is not sufficient for the naturalization (among other things) of the idea of system-detectable error (Sect. 6.3.4). In the part of the book devoted to teleosemantics, I show how the relationship between representation and action can be captured in a naturalistic way.

6.3.1 Davidson's Assumptions

Davidson's conception of meaning as truth conditions assumes a weaker version of the schema:

(P2) s means that p is true only if s is true if and only if p is true.

This means that although meaning is not exhausted by truth conditions, these conditions can be used to describe it. In this version of the theory, the bearers of truth are utterances, and circumstances of the interpreted utterance, such as time or place, are taken into account [6]. The T-schema takes on a modified form:

(T') U considers the sentence s to be true at time t if and only if p is true.

Davidson notes that although each of the three elements separately: theory of action, theory of beliefs and desires, and theory of interpretation, is problematic, they can be used in conjunction [2]. However, he does not reject the assumptions of the extensional theory but rather supplements them. He believes that a unified theory of meaning and action should be created, not treating any of the three elements as basic, but assuming that these elements are interdependent.

This interdependence, as Davidson vividly describes, is akin to triangulation. Just as the position of a point in space can be determined by measuring the angle between it and two known points and measuring the distance from them, so too can the meaning of an utterance be determined by referring to other known parameters of it. The constitutive elements of triangulation in communication are three interdependent relationships: the relationship between the speaker's utterance and the message recipient, the user's relationship with the world, and the message recipient's relationship with the world [10]. This idea can be expressed in terms of mental representations, saying that the content of representations is the resultant of both communicating users' relationships with each other, as well as their relationships with the world.

The theory of interpretation (governed by the principle of charity—see: Chap. 2) takes into account not only utterances and beliefs and desires (degrees of beliefs and desires), but also the communication of rational agents. A satisfactory naturalistic theory should answer the question of why representations have such semantics and not others.

Thus, according to Davidson, an essential factor in determining the content of representations is the communication of rational beings. Davidson introduces a concept to describe this kind of communication, the concept of common ground, which arises between the user and the recipient due to the assignment of propositional attitudes. The content of representations—as Tyler Burge puts it following Davidson [11]—(1) is always the result of social construction, and 2) requires the possession of knowledge by both the interpreter and the user.

Davidson considers animals to be irrational beings that are not capable of acting. I have previously rejected his argument against the rationality of animals,[5] and also argued for mental representations at the subpersonal level (Chap. 2), therefore I modify his thesis about action as communication to also describe the functioning of animals.

6.3.2 *Representational System-Detectable Error*

Davidson points out how crucial it is for a cognitive system to be able to detect an error:

> So far as I can see, no account of error that depends on the classifications we find most natural, and counts what deviates from such as error, will get at the essence of error, which is that the creature itself must be able to recognize error. [12, 13: 141]

He then relates this to veracity:

> A creature that has a concept knows that the concept applies to things independently of what it [itself - K.B.] believes. A creature that cannot entertain the thought that it may be wrong has no concepts, no thoughts. To this extent, the possibility of thought depends on the idea of objective truth, of there being a way things are which is not up to us. [12, 13: 141]

Therefore, the ability of a cognitive system to recognize errors indicates the possibility of correct representation. Both the ability to represent correctly and the ability to represent erroneously are necessary for having representations. By asserting that the possibility of representing depends on the idea of objective truth, Davidson places truth and falsity in an asymmetrical relationship, where truth plays a distinguished role. This is consistent with my key desideratum concerning the role of error in representing:

II.3.3. An adequate theory of representational content takes into account the possibility of a cognitive system recognizing the adequacy of representations.

The distinguished role of truth is also the motivation behind the formulation of the principle of charity (see Chap. 2.).

6.3.3 *Objections*

Davidson recognizes the difficulties of the concept of meaning as conditions of truth and formulates objections to it. Elsewhere, he points out that the indeterminacy of meaning arises during the attribution of beliefs:

---

[5] In Chap. 2, I demonstrate that Davidson does not convincingly justify the claim that ascribing a propositional attitude requires having the concept of belief. I defend a weaker assumption, according to which having a belief (rather than its concept) is sufficient for being ascribed a propositional attitude.

> But our ability to identify, and distinguish among, the propositions an agent entertains is not to be separated from our ability to understand what he says. We generally find out exactly what someone wants, prefers, or believes only by interpreting his speech. [1: 4]

The content of representation is determined only as a result of communication. It seems that this has two undesirable consequences for the theory of representation: (a) it does not allow for the existence of representations that do not require the interpretation of other subjects, but only a relation to the world (and such include, for example, the entire class of perceptual representations), (b) the thesis of social determination of content does not solve the problem of naturalizing intentionality.

According to the NTCS (Naturalized Truth-Conditional Semantics) theory, the content of representation always emerges as a result of agreement with others in the communication process. However, representations do not always require communication. Prima facie, it seems that many individual representations at the personal level do not require communication. For instance, my belief that the weather is nice today does not necessitate the mediation of other people, and the representational relation can be described as one occurring between me and the world. This natural and commonsensical assumption about the individual nature of representations can be challenged by pointing out that representations have a linguistic character, and language is inherently social. However, this argument is not very convincing due to the existence of representations such as cognitive maps, which the semantics of truth conditions captures as distinct from linguistic ones. A naturalizing advocate of Davidson's philosophy might suggest considering cognitive operations as communication at the subpersonal level. Memory can be conceptualized in communicative terms.[6] Thus, the concept of meaning as truth conditions can be applied to cognitive maps, which are not necessarily used by the entire organism but also by its internal mechanisms (as I will show below).

Nevertheless, NTCS cannot resolve the problem of intentionality. I find the project's assumption convincing, according to which truth conditions, without considering contextual factors, do not exhaust meaning. Including additional factors in the triangulation process, such as preferences and action, makes the problem of naturalizing meaning take on a more elusive form. The problem does not disappear by considering additional factors. It remains because it is unclear why a given representation means this and not something else. A naturalizing advocate of Davidson could argue that a given representation means something because it is socially recognized as meaning just that. However, it is unclear which social interpretations of utterances are correct and which are erroneous. To speak of correctness, one would have to acknowledge that meaning is not a matter of arbitrary interpretations, which means that it must depend on something. Yet, it is unclear what that something is. A precise account of the relationship between meaning and action is not provided at all.

---

[6] Godfrey-Smith suggests that memory can be described as sending a signal (information) over time [14]. Taking this suggestion, I developed a communicative account of memory that conceptualizes it as a network of information transfer spanning both internal cognitive processes and external social interactions [15].

6.3.4 *Application of the Conception to Examples from Mind and Brain Sciences*

Rombix

Let us now imagine a subject learning a new concept, ROMBIX. The subject says, "This is a rombix," upon seeing a correctly chosen picture, and "This is not a rombix," in the case of an incorrectly chosen picture. Let us now compare sentences TE1 and TE2 from the previous section:

"The statement 'This is a rombix' is true if and only if a rombix is a red diamond."

"The statement 'This is a rombix' is true if and only if a rombix is a yellow triangle or a rombix is a red diamond."

Davidson's strategy dictates treating the T-schema as:

(T′) U considers the sentence s to be true at time t if and only if p

A convenient tool for generating new hypotheses about meaning. By using the T-schema, the subject adjusts their hypothesis about the meanings of the experimenter's words, who classifies the sentence about the picture as true in the case of TE1 and as false in the case of TE2. The subject verifies their initial hypothesis about the meaning of the word "rombix" when the experimenter indicates that TE2 is false.

The principle of charity, according to which the subject assumes that the experimenter mostly speaks the truth, is applicable in the current experimental situation. It allows for the rejection of some interpretations consistent with the behavioral testimony concerning the subject's behavior. According to the principle of charity, the simplest hypothesis about the name "rombix" is preferred. This distinguishes Davidson's conception from the purely extensional concept of truth conditions discussed in the first part of the chapter, as it does not reject such hypotheses. However, this does not change the fact that Davidson's conception does not show why our words mean anything at all.

Cognitive Maps

Davidson rejects the possibility of using intentional vocabulary to describe animal behavior. His arguments for privileging language are unconvincing, as I demonstrate in Chap. 2. Davidson's conception can be related to cognitive maps or other structural representations, as the semantics of satisfaction conditions can be applied to various types of representations. In Davidson's conception, one could directly utilize the semantics of cognitive maps axiomatized by Rescorla, which I discussed in the previous part of the chapter. Then the question arises whether Davidson's approach would contribute anything new.

According to Davidson, meaning is the result of various factors in triangulation. Triangulation includes, among other things, communicative interactions—however, there is a problem because cognitive maps are not communicated to other rats. The application of the strategy of mutual understanding of meaning between the sender and receiver of the signal (message) at the personal level is highly questionable, as meaning arises in the rat's interaction with the environment, not with another organism.

NTCS does not address how meaning is constituted in interaction with the environment. However, since there is substantial evidence that the application of representations in communication fits the description of memory functioning [16, 17–19], NTCS can be applied to describe the functioning of memory at the subpersonal level. In this case, one should consider which systems use cognitive maps, for example, in planning movement. This would allow the researcher to clarify hypotheses about the content of these maps. Hypotheses about content that would not be utilized by neuronal circuits would be rejected. Memory here is not treated statically, as reflecting reality, but as a repository of messages received and actively processed by the organism's subpersonal systems.

The project of conceptualizing meaning as truth conditions does not show how a creature (in this case, a rat) can be mistaken. However, the possibility of error detection is a crucial assumption of the project and links representation to action: in the event of a mistake, the creature should strive for an adequate representation, as it is beneficial for it. This remains consistent with the assumption of instrumental rationality as the basis for decision theory. Analogous to cognitive maps, mental representations can be treated as a kind of communicative signals (even though—at least according to Kosslyn—they do not have a linguistic nature).

It is unclear why representations have such and not other truth conditions, and this problem is not resolved by NTCS. Nevertheless, conceptualizing content in terms of truth conditions or more broadly—satisfaction, understood not abstractly but within the framework of communication, can serve to refine scientific hypotheses concerning various communication systems. The requirement to define the truth conditions of representations, that is, to determine when they correspond to reality, seems justified. In other words, NTCS requires that descriptive representations be conceived as truth conditions, which partially meets my desideratum about content (II.2.2). However, it does not fully satisfy it, as it does not address why something can bear content.

6.4 Summary

I will gather the desiderata that the concepts discussed in this chapter satisfy as truth conditions. They fulfill the desideratum II.3.1 about productivity, as well as desideratum I.2 about universality. An example of non-linguistic phenomena that the discussed concepts capture are structural representations, such as a cognitive map. The NTCS theory, inspired by Davidson's project, also meets the desideratum about the multi-level applicability.

The general direction of the project aligns with part of my desiderata for a satisfactory conception of representation. This theory, unlike informational-causal semantics, does not exclude the possibility of representational error, thus fulfilling the first desideratum concerning error:

II.3.2. A satisfactory theory of representational content accounts for the possibility of misrepresentation.

Davidson's remarks on mental representations provide justification for the validity of the second desideratum concerning error:

II.3.3 A satisfactory theory of representational content accounts for the cognitive system's ability to recognize the adequacy of representations.

Unfortunately, the discussed concepts do not meet the desideratum:

II.2.2. demonstrates why a representation can bear content,

thereby failing to solve Hutto and Myin's hard problem of content.

Concepts of truth conditions do not lead to the naturalization of the concept of truth but at best provide some clues as to how truth might be naturalized. Davidson points to pragmatic and contextual elements, such as the goals and preferences of cognitive systems operating in an environment, which are constitutive for the content of mental representation.

6.5 Conclusions: Non-teleosemantic Externalism

In Chap. 5 and in this chapter, I analyzed purely extensional concepts of mental representation: in Chap. 5, I characterized informational-causal semantics, and in this chapter, the concepts of meaning as truth conditions.

In the conclusion of Chap. 5, I stated that purely extensional causal Wisconsin-style semantics struggles with the disjunction problem and the treatment of representational error. None of the conceptions mentioned in this chapter can meet the task of naturalizing intentionality. Fodor's asymmetrical conception comes closest to solving the disjunction problem, presenting misrepresentation as dependent on correct representation: error is conceived as dependent on correctness, but not vice versa. Fodor rightly points to the asymmetry between erroneous and correct representations, but it is unclear how such asymmetry arises. This allows us to avoid the less credible assumptions typical of other causal concepts, which refer to peculiar normal conditions or expert knowledge. Although Fodor's solution to the disjunction problem is better, there are convincing arguments against his treatment of misrepresentation. The asymmetrical relation also occurs between non-representational objects, such as a pick lock and a true key or toxin and neurotransmitter (as noted by [20]). In other words: Fodor cannot repel the criticism that his concept is too liberal and trivializes the concept of mental representation.

In this chapter, I pointed out that truth conditions are closely related to content. Although Davidson's conception acknowledges the possibility of misrepresentation, it does not provide a naturalistic explanation of how a cognitive system can detect representational error. In other words, Davidson does not meet the desideratum II.3.3 concerning error detectable by the system.

I will now indicate which desiderata are met by the non-teleosemantic externalist conceptions discussed in this part, and which are not. Informational-causal semantics satisfies the desideratum concerning the inclusion of reference (II.2.1), but it has a problem with accounting for the possibility of error (II.3.2 and II.3.3). By not considering truth, purely extensional causal conceptions do not meet the desideratum about error detectable by the system (II.3.3). Thanks to fulfilling the desideratum concerning the content bearer (II.2.2), the purely extensional conception of truth conditions copes with the desideratum about naturalizing intentionality (II.2) somewhat better than internalist conceptions. However, it does not allow for the naturalization of intentionality (thereby not solving the hard problem of content by Hutto and Myin). The conception of meaning as truth conditions (especially in Davidson's version) meets the desideratum II.3.2 concerning the possibility of error. It does not, however, fulfill the desideratum II.3.3 about error detectable by the system, as it does not show how a cognitive system can have the ability to detect the adequacy of its own representations. Thus, neither informational-causal nor truth-conditional conceptions of content are capable of naturalizing intentionality.

References

1. Davidson, D.: Toward a unified theory of meaning and action. Grazer Philosophische Studien **11**(1), 1–12 (1980). https://doi.org/10.1163/18756735-90000093
2. Davidson, D.: Rational animals. Dialectica **36**(4), 317–327 (1982). https://doi.org/10.1111/j.1746-8361.1982.tb01546.x
3. Field, H.: Tarski's theory of truth. J. Philos. **69**(13), 347–375 (1972)
4. Davidson, D.: Inquiries into truth and interpretation. Clarendon Press, Oxford (1984)
5. Barwise, J., Perry, J.: Semantic innocence and uncompromising situations. Midwest Stud. Philos. **6**(1), 387–404 (1981)
6. Davidson, D.: Truth and meaning. Synthese **17**(1), 304–323 (1967). https://doi.org/10.1007/BF00485035
7. Foster, J.A.: Meaning and truth theory. In: McDowell, J.H., Evans, G. (eds.) Truth and meaning: essays in semantics, pp. 1–33. Clarendon Press, Oxford (1976)
8. Harman, G.: (Nonsolipsistic) conceptual role semantics. In: LePore, E. (ed.) New directions in semantics, pp. 55–81. Academic Press, London (1987)
9. Rescorla, M.: Cognitive maps and the language of thought. Br. J. Philos. Sci. **60**(2), 377–407 (2009). https://doi.org/10.1093/bjps/axp012
10. Davidson, D.: The folly of trying to define truth. J. Philos. **93**(6), 263–278 (1996). https://doi.org/10.2307/2941075
11. Burge, T.: Origins of objectivity. Oxford (2010)
12. Davidson, D.: What thought requires. In: Problems of rationality, pp. 135–151. Oxford University Press, Oxford (2004)
13. Davidson, D.: The Problem of Objectivity. In: Problems of rationality, pp. 3–19. Oxford University Press, Oxford (2004)
14. Godfrey-Smith, P.: Signals, icons, and beliefs. In: Ryder, D., Kingsbury, J., Williford, K. (eds.), Millikan and her critics, pp. 41–62. Wiley-Blackwell, Malden (2012)
15. Bielecka, K.: Confabulations in Korsakoff's syndrome: defending an error-based account. Erkenntnis (2024). https://doi.org/10.1007/s10670-024-00857-w
16. Gallistel, C.R.: The Nature of learning and the functional architecture of the brain. In: Jing, Q., et al. (eds.) Psychological science around the world. Proceedings of the 28th International Congress of Psychology, vol. 1, pp. 63–71. Psychology Press, Brighton (2006)
17. Gallistel, C.R., King, A.P.: Memory and the computational brain. Wiley-Blackwell, Oxford (2009)

18. Millikan, R.G.: Language, thought, and other biological categories: new foundations for realism. MIT Press, Cambridge (1984)
19. Millikan, R.G.: Varieties of meaning: the 2002 Jean Nicod lectures. MIT Press, Cambridge (2004)
20. Bickhard, M.H.: The interactivist model. Synthese **166**(3), 547–591 (2009). https://doi.org/10.1007/s11229-008-9375-x

Part II
Teleosemantic Accounts of Representation

Teleosemantics frames representations in terms of teleological function. The concept of teleological function is sometimes understood in a manner akin to functions found in evolutionary biology, and partly in the manner of functions found in systems biology.[1]

The naturalistic concepts discussed so far equate representational error with the absence of representation, thereby failing to account for the content of misrepresentations or for misrepresentations themselves. An exception is Fodor's concept of asymmetric dependency, which, however, has certain shortcomings. Most importantly, it does not elucidate what constitutes misrepresentation. Teleosemantics, on the other hand, by understanding concepts such as representing, misrepresenting, and non-representing in terms of function, dysfunction, and lack of function, respectively, is much more adept at addressing the issue of misrepresentation than other naturalistic concepts, which reduced misrepresentation to a lack of function.

Within teleosemantics, there is a debate over what exactly constitutes a biological function. Essentially, a function can be understood in three ways—as a need-based function [1, 2], historical (etiological) function [3, 4, 5, 6, 7, 8], or synchronic function [9, 10], including the self-maintenance synchronic function [11, 12, 13, 14, 15]. The synchronic approach in teleosemantics is not typical [16, 17], as it consistently identifies features as functional—whether in terms of successful action or the survival of the organism. Teleosemantics is considered the most mature approach to representation [18]. Evidence of this is that anti-representationalist critiques focus precisely on teleosemantics [19, 20].

The simplest is the concept of need-based function, defined in terms of an organism's basic needs. Evolutionary biologists, on the other hand, understand "function" as the role that a given organ or organism plays (or more precisely, can play) by

---

[1] The literature on the concepts of function in biology is extensive, e.g. Davies, who clearly distinguishes systemic from teleological functions, refers to functions defined in the pattern of those found in systemic biology as hybrid, because they combine a systemic approach with a teleological one. Unlike Davies, who is skeptical of the project of naturalizing normativity, I do not abandon the teleological understanding of functions. Therefore, I do not claim—as he does—that the systemic conception of functions should be conceptually separated from the teleological.

virtue of natural selection. Thus, the stomach was selected for[2], adapted, and in this sense designed for the digestion of food within the organism. Systems biologists, in turn, emphasize the survival and self-preservation of the organism.

A teleological biological function differs from the classical concept of synchronic function according to Cummins, which states that:

x functions as a φ in *s* (or: the function of *x* in *s* is to φ) relative to an analytical account *A* of *s*'s capacity to ψ just in case *x* is capable of φ-ing in *s* and *A* appropriately and adequately accounts for *s*'s capacity to ψ by, in part, appealing to the capacity of *x* to φ in *s* [21, 762].

For instance, the stomach is deemed to have the function of digesting proteins because the digestion of proteins by the stomach is crucial for the functioning of the digestive system that transports nutrients to other organs of the body. An explanation of protein digestion in terms of the stomach's parts and their operations is correct because, among other reasons, a stomach with such parts (e.g., a part secreting hydrochloric acid) is capable of digesting proteins. In short, function in Cummins' synchronic approach is a certain ability (or disposition) of a complex system whose parts are causally interconnected.

Dispositions can be attributed relatively freely, which allows for a liberal application of the concept of function in the sense of Cummins. Millikan regards this as the Achilles' heel of this approach, emphasizing that such a defined concept is not normative. In line with Cummins' definition, we might consider, for example, that the function of the stomach is to cause reflux, whereas intuitively we would deem this a dysfunction. Although the synchronic function in the sense of Cummins differs from the teleological concept of function and is usually not teleological (as emphasized by Paul Davies [17]), it forms the basis of some teleological conceptions of function, namely synchronic biological functions. These are competitive with historical teleological functions.

It is therefore worth comparing Cummins' conception with the historical (etiological) conception. The origins of the historical approach to function in teleosemantics can be found in [33], who posits that

The function of *X* is *Z* means

(2) (*a*) *X* is there because it does *Z*,

(*b*) *Z* is a consequence (or result) of *X*'s being there [33, 161].

For example, the function of a mole's claws (X) is to dig through soil (Z), when Digging through soil (Z) is a consequence of the existence of the claws (X).

The claws exist [or existed, or will exist], because they contribute to digging through soil (Z).

Condition 2 is fundamental in the etiological concept of function because it pertains to the relationship of function to the mechanism of selection. I will pause for a moment on the more controversial Condition 1. It is rejected by later proponents of

---

[2] The concept of selection for a trait was introduced to the philosophy of biology by [22]. More on this subject can be found in the chapter discussing Millikan's theory.

teleosemantics, as it requires attributing to X the effect Z of X's action, which would imply that a biological trait cannot have a function if X does not produce such an effect. For instance, the function of a mole's claws is to dig soil only if they actually cause such effects. Fulfilling Condition 1 precludes distinguishing dysfunction from the absence of function: in the case of dysfunction, it is known that the function of a mole's claws is to dig soil, even though the mole is unable to do so for some reason (e.g., its claws have been damaged by a file by a malicious child); in the second case, however, we would say that the mole's claws do not possess such a function at all.

Synchronic biological function refers to the concept of function in the sense of Robert Cummins. It is thus a certain capacity of a complex system, whose parts are causally interconnected. Advocates of synchronic biological function combine Cummins' idea either with the disposition for efficient functioning[3] [23, 24], or with self-preservation or self-maintenance of the system [12, 25, 26]. A characteristic feature of this function, regardless of its type, is its teleological nature. This connects systemic biological function with etiological function.

The function of representation in teleosemantics is a biological function of a special kind. Reference to biological purposiveness allows for a naturalistic description of intentionality, as well as the normativity attributed to representations. It is therefore worthwhile to examine how teleosemantics fulfills the meta-desiderata and desiderata defended in this book and why it fails to meet a few of them. Some of these are contained within the teleosemantic thesis itself, while others arise from assumptions shared with other naturalistic conceptions. However, it should be emphasized that due to their biological nature, teleosemantic theories allow for a better justification of certain desiderata.

Meta-desideratum I concerning the relationship between representation theory and scientific research is shared by the naturalistic theories under discussion. The explanatory desideratum I.1 is also fulfilled in teleosemantics, as representations are particularly applied to the description, explanation, prediction, and control of cognitive processes and behaviors. Teleosemantic theories, being biological, meet desideratum I.2 regarding universality, and are particularly applicable to the description of non-linguistic phenomena occurring among animals.

Teleosemantics considers linguistic phenomena as deriving from a broader group of non-linguistic phenomena. Thus, Millikan's biosemantics (as the author refers to her theory), as the name suggests, is primarily a biological concept of communication, encompassing, among other things, linguistic behaviors; in Dretske's case, the discussion is about indicators, which constitute a basic level of representation, including both linguistic and non-linguistic behaviors. Since the goal of teleosemantics, like many functionalist conceptions, is to describe the cognitive architecture of the system, it provides tools for describing both personal and subpersonal phenomena, thus meeting the meta-desideratum I.3 concerning the multi-level applicability. Moreover, by describing in functional terms how one organ depends on another, teleosemantics can explain the purposefulness (in a biological sense) of

[3] Anderson, Rosenberg, and Bickhard employ the concept of action in the sense of the whole organism's activity [23, 24].

organs within the functioning of the entire organism. Dretske's concept emphasizes primarily the needs of the individual organism. In Millikan's theory, where individual and social traits are equally important, the concept of representation can be applied at both levels to describe both individual and social phenomena. As a result, Millikan's theory fully accounts for the diversity of cognitive processes in the biological world. Furthermore, her approach does not exclude social representations (an interesting issue, although its development goes beyond the scope of this work).

Meta-desideratum II concerning the relationship between the theory of representation and everyday practice requires the explication of the pre-existing concept of representation. Teleosemantics justifies why descriptions of cognitive systems' operations in intentional terms are correct. Assuming that the conception of the intentional stance can be treated as a correct philosophical analysis of everyday practice (see Chap. 1) and that the intentional stance must be teleosemantically justified, it is easy to show why a cathedral does not have mental representations, while Phineas Gage does. The former lacks the function of representing, while the latter possesses it.

Teleosemantics fully satisfies—unlike other naturalistic conceptions discussed in the book—the extremely important desideratum II.2 on the naturalization of intentionality. It ensures the relationship of representation to reference (desideratum II.2.1) and indicates why a representation can bear content (desideratum II.2.2). The inclusion of conditions of satisfaction is closely related to the possibility of misrepresentation (desideratum II.3.2). This desideratum is embedded in the thesis of teleosemantics and is related to the concept of representational dysfunction. It is worth emphasizing that all the teleosemantic conceptions discussed meet this requirement.

I will now introduce the frog example, which will facilitate the exposition of ideas contained in teleosemantic conceptions. I will not discuss the previously introduced examples of the snap trap, cognitive map, and mental imagery for each teleosemantic view separately, as they are very similarly addressed in them. Therefore, I will discuss them collectively at the end of the section devoted to teleosemantics. The frog example is also widely discussed among teleosemanticists, and they differ in how they describe frog representations [27, 28]. Here is what the empirical data tells us about the frog:

When a fly enters a frog's field of view, some of the frog's retinal neurons are activated and send a signal to its brain, causing it to go into a state of type N. This state, in turn, causes a characteristic feeding response: the frog's snapping for the fly and, in case of success, swallowing and digesting it [28, 484].

Flies are food for frogs, towards which a hungry amphibian extends its tongue. However, a hungry frog may also extend its tongue towards any object that is small, dark and moves on the frog's horizon (within the field of vision and in the right place). Such objects include other dark insects, pieces of bark falling from trees, and black buttons tossed by malicious experimenters, which do not normally occur in the frog's natural environment.

The frog example has sparked numerous controversies among philosophers, psychologists, and cognitive scientists. The greatest challenge arises from the question of the content of the frog's representation—is it a) a fly at location m, b) a small, dark, moving object at location m [28, 29], or c) frog's prey at location m [3]? It is less often doubted whether the frog actually represents [28, 30], as well as whether it can misrepresent [31]. For the purposes of this book, I assume that frogs can represent and misrepresent. This is supported by numerous empirical facts that indicate amphibians possess cognitive and motivational abilities, which allow us to attribute to them the capacity for representation [28]. These are basic abilities of identification, reidentification, and distinguishing certain features or objects (tracking mechanism[4]; integration of multiple senses) and susceptibility to change through learning [32].

According to teleosemanticists, attributing representation to the frog primarily aims to explain the organism's effectiveness in action, that is, its success in flexible adaptation to changing environmental conditions. Since the frog's fundamental biological goal is survival, its behavior is oriented, among other things, toward hunting for prey in the form of flies. The frog's activity is concluded successfully if it can capture the encountered fly. Since the frog has the ability to distinguish edible flies from other inedible objects (let's assume these are bee-bees), it can be attributed to the capacity for representing (edible flies as distinct from inedible non-flies). Thus, the frog's current activity resulting from representing (the fly) can be concluded successfully, among other things, when it is consistent with reality, when a fly is indeed present in the frog's field of vision.

Given the theme of this work, I will focus on misrepresentation. For example: the frog misrepresents an edible fly when it extends its tongue in the presence of a bee-bee. The frog misrepresents an edible fly when, for instance, it lunges at a bee-bee, which then may harm its digestive system. The frog can be attributed the ability to recognize a mistaken representation if, for example, the bee-bee actually harms its digestive system and another time, thereafter, in its presence, the frog will not lunge at it. This means that the frog is capable of recognizing from the consequences of its actions that it made a mistake and learning how to proceed in the future.

Reference

1. Dretske, F.I.: Knowledge and the Flow of Information. Cambridge University Press, Cambridge (1981)
2. Dretske, F.I.: Misrepresentation. In: Bogdan R. (ed.), Belief: Form, Content, and Function, pp. 17–37. Clarendon Press, England (1986)
3. Millikan, R.G.: Language, Thought, and Other Biological Categories: New Foundations for Realism. Cambridge; MIT Press, USA (1984)
4. Millikan, R.G.: Biofunctions: two paradigms. In: Ariew, A., Cummins, R., Perlman, M. (eds.), Functions: New Essays in the Philosophy of Psychology and Biology, pp. 113–143. Oxford University Press, New York, USA (2002)
5. Dretske, F.I.: Naturalizing The Mind. The MIT Press, USA (1997)

[4] I discuss the tracking mechanism in more detail in Chaps. 7 and 9.

6. Neander, K.: Misrepresenting & malfunctioning. Philosophical Studies **79**(2), 109–141 (1995)
7. Neander, K.: Teleological theories of mental content. In: Zalta, E.N. (ed.), The Stanford Encyclopedia of Philosophy (2012). Retrieved from http://plato.stanford.edu/archives/spr2012/entries/content-teleological/
8. Macdonald, G., Papineau, D.: Teleosemantics: New Philosophical Essays. Oxford University Press, Oxford, New York (2006)
9. Rosenberg, G., Anderson, M.L.: A brief introduction to the guidance theory of representation. In: Forbus, K., Gentner, D., Regier, T., (eds.), Proceedings of the 26th Annual Conference of the Cognitive Science Society, pp. 55–81. Lawrence Erlbaum Associates, Hillsdale (2004).
10. Anderson, M.L., Rosenberg, G.: Content and action: the guidance theory of representation. J. Mind Behav. **29**(1–2), 55–86 (2008)
11. Bickhard, M.H., Terveen, L.: Foundational Issues in Artificial Intelligence and Cognitive Science: Impasse and Solution. North-Holland (1995)
12. Bickhard, M.H.: The interactivist model. Synthese **166**(3), 547–591 (2009). 10.1007/s11229-008-9375-x
13. Bickhard, M.H., Christensen, W.: The Process Dynamics of Normative Function. Monist **85**(1), 3–28 (2002)
14. Campbell, R.J.: The concept of truth. Palgrave Macmillan, New York (2011)
15. Moreno, A.: Biological autonomy: a philosophical and theoretical enquiry. Springer, Dordrecht (2015)
16. Cummins, R.: Functional analysis. J. Philos. **72**(20), 741–765 (1975). 10.1007/978-1-4614-7004-5
17. Davies, P.S.: Norms of Nature: Naturalism and the Nature of Functions. MIT Press, Cambridge (2001)
18. Neander, K.: Teleological theories of mental content. In: Zalta, E.N. (ed.), The Stanford Encyclopedia of Philosophy (2012). Retrieved from http://plato.stanford.edu/archives/spr2012/entries/content-teleological/
19. Chemero, A.: Anti-representationalism and the dynamical stance. Philos. Sci. **67**(4), 625–647 (2000)
20. Ramsey, W.M.: Representation Reconsidered. Cambridge University Press, Cambridge (2007). 10.1017/CBO9780511597954
21. Cummins, R.: Functional analysis. J. Philos. **72**(20), 741–765 (1975). 10.1007/978-1-4614-7004-5
22. Sober, E.: From a Biological Point of View: Essays in Evolutionary Philosophy. Cambridge University Press, Cambridge (1984)
23. Rosenberg, G., Anderson, M.L.: A brief introduction to the guidance theory of representation. In: Forbus, K., Gentner, D., Regier, T. (eds.), Proceedings of the 26th Annual Conference of the Cognitive Science Society, pp. 55–81. Lawrence Erlbaum Associates, Hillsdale (2004)
24. Anderson, M.L., Rosenberg, G.: Content and action: the guidance theory of representation. J. Mind Behav. **29**(1–2), 55–86 (2008)
25. Bickhard, M.H., Christensen, W.: The process dynamics of normative function. Monist **85**(1), 3–28 (2002)

26. Bickhard, M.H., Richie, M.D.: On the nature of representation. New York (1983)
27. Lettvin, J.Y., Humberto, H.R., Maturana, R., McCulloch, W.S., Pitts, W.H.: What the frog's eye tells the frog's brain. Pro. IRE **47**(11), 1940–1951 (1959)
28. Neander, K.: Misrepresenting & malfunctioning. Philos. Stud. **79**(2), 109–141 (1995)
29. Schulte, P.: How frogs see the world: putting millikan's teleosemantics to the test. Philosophia **40**(3), 483–496 (2012). 10.1007/s11406-011-9358-x
30. Burge, T.: Origins of Objectivity. Oxford (2010)
31. Miłkowski, M.: Explaining the Computational Mind. MIT Press, Cambridge (2013)
32. Schmajuk, N.A., Segura, E.T., Reboreda, J.C.: Appetitive conditioning and discriminatory learning in toads. Behav. Neural Biol. **28**(4), 392–397 (1980). 10.1016/S0163-1047(80)91698-2
33. Wright, L.: Functions. Philosophical Review, **82**(2), 139–168 (1973)

Chapter 7
Dretske's Account of Misrepresentation as Dysfunction

The structure of the chapter is as follows. First, I will describe Dretske's motivations (Sect. 7.1), then his conception of the representational function as a function of indicating (Sect. 7.2), and explain how he analyzes representational error in terms of dysfunction (Sect. 7.3). I will show that—different from the causal-informational approach—a functional account of the indication relation allows Dretske to tackle the hard problem of content (Sect. 7.2). I will highlight the advantages and disadvantages of his theory, particularly those concerning the treatment of misrepresentation. Next, I will apply Dretske's account of misrepresentation to the example of the frog introduced in the introduction (Sect. 7.4). Finally, I will demonstrate how Dretske deals with the disjunction problem (Sect. 7.5).

7.1 Introduction to Dretske's Theory. Motivations

Dretske's mature concept is an example of informational semantics, the foundations of which are described in his earlier book *Knowledge and the Flow of Information* [1]. Dretske provided an influential account of semantic information, which was later simplified and replaced by the concept of natural meaning in his work on mental representation.

Dretske was the first to so clearly emphasize the role of error in representation. He even dedicated a separate paper, *Misrepresentation* (Dretske 1984) to analyze primarily what error is. It suffices to quote the first sentence of the article:

> Epistemology is concerned with knowledge: how do we manage to get things right? There is a deeper question: how do we manage to get things wrong? How is it possible for physical systems to *misrepresent* the state of their surroundings?. [2: 17]

Dretske starts from common-sense intuitions about the role of error in the conception of mental representation and tries to follow them throughout the article. The fundamental idea to which he refers not only in *Misrepresentation* but also in his

K. Bielecka, *What is Misrepresentation?*, Studies in Applied Philosophy, Epistemology and Rational Ethics 71, https://doi.org/10.1007/978-3-031-84375-4_7

entire body of work is based on the observation that cognitive systems are characterized by the ability to learn, that is, acquiring knowledge and concepts about the world around them. Since learning is a process that involves continuous erring and self-correction, thus adapting one's behavior to environmental conditions, the system's ability to misrepresent is a necessary condition for learning. Error is a signal that the system should change its behavior. Dretske's theory shows how a rational organism can falsely represent objects and their properties in the world around it.

The ability to learn from mistakes requires access to natural meaning, which, according to Dretske, is possessed only by natural cognitive systems. Dretske, like Searle, attributes mental processes only to biological creatures endowed with genuine intentionality. In this concept of representation, there is room to describe both linguistic and non-linguistic phenomena (e.g., perceptual) of humans and animals alike (consistent with the desideratum I.2 about universality).[1]

7.2 The Conception of the Function of Representation as a Function of Indicating

According to Dretske, the function of representation is the function of indicating [2].[2] Dretske analyzes representation in terms of functional meaning, based on natural meaning in Grice's sense:

> Grice expresses this point by saying that an occurrence (a tokening of some natural sign) means (...) that P only if P. He contrasts this sense of meaning with non-natural meaning where a sign can mean that P even though P is false. [2: 20]

A classic example is thunder as a sign of a storm: thunder always informs us of a storm and in this sense always represents a storm if it occurs; however, it does not represent a storm if thunder does not occur.

Dretske defines functional meaning as follows:

> (M_f) *d*'s being *G* means$_f$ that *w* is F = *d*'s function is to indicate the condition of *w*, and the way it performs this function is, in part, by indicating that *w* is *F* by its (*d*'s) being *G*. [2: 22]

This is illustrated as follows: a thermometer (d) has a column height (G), which means that the air (w) has a temperature of 20 °C (F). Similarly, for sensory receptors:

> Just what the various states these receptors mean$_f$ will (in accordance with M_f) be determined by two things: (1) what it is the function of this receptor system to indicate, and (2) the meaning$_n$ of the various states that enable the system to perform this function. [2: 25–26]

[1] Dretske wrote about language quite succinctly. According to him, linguistic behaviors can be described using conventional representations [3: 8].

[2] Initially, Dretske conceptualized function in terms of needs [1, 2] and in his mature account, he employed the concept of historical function [3]. Both biological concepts of function are consistent with the theory of evolution and natural selection. In the need-based function concept, an organism and its parts have a function when they perform certain actions due to biological needs. Functionality in the conception of historical function is the result of natural selection. I will discuss the historical account in more detail by introducing Millikan's account of proper function.

Although Dretske grounds functional meaning on natural meaning, he does not reduce the former to the latter. This distinguishes his conception from informational-causal ones, where there is no room for representational error—an error is immediately reduced to a lack of representation. Natural meaning is determined by regular co-occurrences between events in the world due to laws or probabilistic dependencies. As a result, functional meaning is independent of the arbitrary intentions of the observer, yet it pertains to the functioning of the organism.

The indicating function can be captured in the form of a definition: a cognitive system (or its part) has the function of representing when the system (or its part) has the function of indicating that w is f by virtue of d being g.

Allow me to digress briefly here regarding the definition of the function of representation in terms of an indicator. Robert Cummins and William Ramsey are critical of conceiving representation in terms of the function of indicating, considering it a category mistake [4]. However, their arguments seem problematic because they inaccurately attribute to Dretske a simplistic conception of the indicator's structure.

According to Cummins, the structure of indicator signals in no way resembles the structure of representation [4: 367]. He argues that the difference in the structure of indicator signals stems from the fact that the relation of representation (mapping) has a different nature from the nature of the indicating relation. The indicating relation is characterized by transitivity, which allows anything to be made an indicator of anything else. Transitivity thus implies that indicator signals are arbitrary, precluding systematicity: systematic transformations of structure cannot systematically change content. From the very nature of the indicating relation, it follows that the structural properties of the indicator—unlike those of representation—do not carry semantic information. As Cummins writes:

> Indicator signals demonstrate that their targets are there, but are silent about what they are like. [4: 367]

While it is true that the indicating relation is transitive, Dretske understands this relation quite differently than Cummins accuses him of. It seems that what Cummins considers the structure of the representing relation is precisely what Dretske regards as the structure of the indicating relation, which in his conception always not only indicates an object but also attributes certain properties to it. Cummins' argument would hold if representation were conceived in terms of the simplest indicators, whose indicator signals lack structure. However, these are practically non-existent cases, especially in the field of mental representation. Typical examples of mental representations are precisely indicators with complex structure, that is, S-representations in Cummins' sense [4, 5]. Contrary to Ramsey [6], most perceptual representations are indicators with a rich structure that maintain a structural resemblance to their referents [7].

Thus, the charge of a category mistake in defining representation by the relation of indication is to some extent misplaced, as it arose from poorly chosen examples by Dretske, such as a speedometer or thermostat; Dretske himself would likely not want the occurrence of the relation of indication to exclude more complex cases of representation, described in terms of structural similarity relations. Importantly, not all teleosemantic approaches rely on the notion of indication (e.g., Millikan's theory), would not be susceptible to the charge of transitivity of this relation.

The function of indication is based on the concept of an indicator receiving information. In complex systems, an indicator has its role in their architecture, whereby the indicated information elicits specific behavior. For instance, a heat receptor on the skin informs the organism how to maintain a constant body temperature by indicating that the body temperature (d) is 36 °C (g), which means that the air temperature (w) is 38 °C (f).[3] Importantly, Dretske believes that mental representations additionally require conceptual abilities. While a thermoreceptor indeed has the function of indicating body heat, it cannot have mental representations, as it lacks concepts.

Nevertheless, indicators are already bearers of content, as analogously to predicates, they assign certain properties to correlated objects or their features. Since what has the function of indication can also have a dysfunction of indication, what is to serve as an indicator (e.g., a thermoreceptor) can attribute false properties to correlated objects or their features. Typically, thermoreceptors accurately indicate the surrounding temperature, thus fulfilling the function of indication, but they can also falsely indicate (have a dysfunction of indication) when they do not fulfill the function of indication. What has the function of indication also has satisfaction conditions–these are namely the conditions under which the function of indication is realized. Dretske thus naturalizes the satisfaction conditions and thereby resolves—contrary to the objections of Hutto and Myin—the hard problem of content [8].

Dretske utilizes the indication function to describe the behavior of the cognitive system at both the personal and subpersonal levels, thereby fulfilling desideratum I.3 concerning multilevel analysis. Let's assume that instance C is an occurrence (of a physical condition) of type N, whose correlation with condition type M causes instance C of type N to indicate (carry information) that F is occurring, because the relationship between N and M (established through learning) causes instance N (when the prevailing circumstances—whether motivational or otherwise—are appropriate) to contribute to M. For example, my trip to the refrigerator at 5:00 PM yesterday (C) leads to the extraction of vanilla ice cream (M). My behavior C has meaning—for instance, when it represents my desire to eat ice cream F—because it is an instance of behavior type N, which, due to the properties of indicating (through correlation with F), plays a role in controlling the movements of my body (as the cause of behavior M), by indicating the desire to eat ice cream F—see [9: 215]. We can speak of representing when indicating leads to the regulation of behavior.

Dretske thus naturalizes intentionality in terms of the indication function, which is extrinsic:

> I, myself, favor an information-based theory of meaning wherein the meaning of (primitive) conceptual elements is the information they have the job of carrying (their indicative function) … this makes the meaning of an element depend on its relations - specifically, its correlations - with the environmental conditions about which it is supposed to carry information. On this view meanings are extrinsic. [9: 203–204]

[3] The temperature difference (here chosen arbitrarily, solely for the purpose of illustrating the functioning of an indicator) occurs due to the inaccuracies of thermoreceptors.

He analyzes indication in terms of the covariance between objects (or their features). Neither covariance alone nor covariance strengthened by the concept of normal conditions of its occurrence provide conditions for the correctness of representation (unlike in information-causal concepts). Dretske refers to the normal conditions of proper functioning, that is, indicating. They are objective because they carry information about environmental conditions—it is these conditions, which most often accompany the function of representing, that are the normal conditions of representation. Thus, the function of a thermometer, which has been properly calibrated to measure temperature in Celsius, is to indicate a specific temperature under specific environmental conditions. A well-calibrated thermometer may malfunction if it is in conditions where, for example, the pressure has changed, to which the thermometer's scale is not adapted. Therefore, normal conditions are probabilistic dependencies—for instance, a thermometer has the function of indicating temperature because it is properly constructed, and its physical properties are dependent on the environment.

7.3 Mental Representation

Representations form a system, within which individual representations are interconnected. As Dretske writes:

> The states by means of which a representational system performs its informational functions have a structure that enables them to acquire functions without explicitly acquiring them. Once we assign "12" to a certain position of the clock hands, all the remaining indicator states (positions of the clock hands) receive an "implied" indicator function. [3: 21]

Individual representations or representational states have an indication function, defined with respect to the natural meaning of the individual states of the organism. Thus, one can speak of a kind of systematicity, as a system with representations has indicators among which certain systematic relations occur. Nevertheless, Dretske nowhere includes compositionality as a condition for indicating, which complicates the fulfillment of desideratum II.3.1 on productivity.

Dretske identifies three constitutive features of representations in cognitive systems, two of which are exclusively constitutive for mental representation. These features are: the capacity for misrepresentation, individual learning of the cognitive system, and the availability of representations in the system's consciousness. The capacity for misrepresentation is a more general feature that distinguishes representation from processes of other kinds. The remaining features: the characteristic of individual learning of the cognitive system and the availability of representations in the system's consciousness, pertain solely to mental representations. Additionally, Dretske associates the origin of representational content with mental representation.

Dretske characterizes the representing system as having the function of perceptual representation. However, the function of conceptual representation is attributed only to those cognitive systems that have the function of conceptual representation.

The function of conceptual representation is based on the function of perceptual representation:

> ... those creatures who have separate detector systems for each nutrient they need and inflexible (not susceptible to change by learning) responses to each, have, on this account, no experience of the food their detectors detect. Since they have no cognitive system for the detectors to service, no way of recalibrating in terms of individual needs and circumstances, their representations of food do not qualify as experiences of food. [3: 173]

According to Dretske, perceptual representations can form the basis of sensory experience, provided that the organism is equipped with a cognitive system that allows for recalibration. This implies that only such organisms that are capable of experience, i.e., those that are cognitively flexible, possess concepts. Only such organisms create mental representations. In other words: the flexibility of the cognitive system is a necessary condition for an organism to have mental representations and, consequently, to be capable of experience.

Here, we must distinguish three interrelated issues: (1) cognitive architecture: positing an organism's cognitive system through which it can flexibly understand reality; (2) defining the nature of mental representations as distinct from perceptual ones; (3) the condition of learning, as well as the connection of learning with the nature of mental representations.

The first issue concerns the architecture of the cognitive system as a condition for mental representation. This condition is common to all teleosemantic concepts discussed; however, in Dretske's version, it requires further clarification. In his teleosemantics, it is only known that the cognitive system serves "to operate detectors" and appears in organisms with a sufficiently complex cognitive architecture.

The nature of mental representations and learning are closely related. According to Dretske, individual learning of the cognitive system differentiates mental representations from perceptual ones. Thus, from the competence point of view, representational systems are those that have the ability to learn individually.

In Dretske's account, mental representations are a particular subset of perceptual representations that the cognitive system can utilize, as a result of learning, to control and regulate its behavior [3: 19]. Dretske [10] also allows for the existence of innate representations, acquired through trial-and-error learning or as a result of natural selection, which he believes is analogous to this type of learning.

The function of representation is acquired through the cognitive system's individual learning from errors. Dretske refers to a particular case of this kind of learning—individual learning through reinforcement. Instances of learning that constitute mental representation are to be distinguished from processes based on innate characteristics, which do not constitute mental representation, such as phylogenetic inheritance processes. However, this distinction seems arbitrary, as learning in biology is not strictly separated from processes based on innate characteristics. Therefore, the distinction between mental representations and perceptual ones, based on individual learning, is unfounded. There is equally significant non-individual learning, and the processes of natural selection, as characterized by Dretske, are described in a way too similar to learning to be considered irrelevant for mental representation.

Dretske, in essence, assumes that individual learning is the hallmark that sets mental representation apart from perceptual representation. Yet, this assumption is undermined by the fact that there is no behavioral evidence to rule out the possibility that other cognitive systems may also contribute to an individual's mental states, making it unclear whether individual learning is a unique characteristic of mental representation [11].

The second distinguishing feature of mental representation related to learning is access to the content of one's own representations. Mental representations are used for the control and regulation of behavior. Only such information available from the senses that can be further subjected to conceptual processing becomes the content of representations. The cognitive system has mental representations when it has access to them through concepts that enable the processing of information received during learning. This means that such processing requires the flexibility mentioned by Dretske (in the above-quoted passage). Here, "flexibility" is understood as "susceptibility to learning," that is, the susceptibility or capacity for re-calibration of behavior ("re-scaling"). According to Dretske, only organisms endowed with concepts are capable of such understood flexibility.

Flexibility further distinguishes individual learning from processes based on innate characteristics. *Prima facie*, an individual cognitive system that learns in a flexible manner is distinct from those systems (artificial or natural) or their parts that have been designed in some way or that involve processes analogous to natural selection. However, flexibility is not a defining feature of individual cognitive system learning. Bacterial populations evolve very rapidly, responding flexibly, for example, to the emergence of antibiotics and becoming resistant to them. Even more so, it is unclear how flexibility could be a feature exclusively attributed to organisms capable of conceptual operations. Therefore, flexibility cannot be a strict hallmark of mental representations.

7.3.1 Misrepresentation

Dretske identifies the erroneousness of representation with indication dysfunction. Indication dysfunction is not the same as the absence of the indicating function. Therefore, if an organism (or its part) has the indicating function, it necessarily has indication dysfunction under certain circumstances (this occurs when the normal conditions for the indicating function are not met). An organism (or its part) has indication dysfunction when its representational state does not naturally mean what it indicates.

It is worth noting that Dretske does not distinguish between representational dysfunction and dysfunction that merely contributes to it. This is not surprising, given that the functioning of the representational system is influenced by the functioning of the entire organism. The circumstances in which indication dysfunction occurs result from abnormalities in the organism itself or its representational system. These circumstances thus include both physical abnormalities in the organism's structure and abnormalities in the flow of information. For example, a predator has indication dysfunction in identifying its prey when its vision fails. The fact that the representational system is also influenced by the physical construction of the organism stems from the teleological nature of the representational function—the basis for such a connection is content-based, teleological-informational.

More importantly, in Dretske's account, the possibility of representational error is not what distinguishes mental representation; it is not misrepresentation that sets the boundary between mental and non-mental representations, but rather learning through trial and error. Misrepresentation can also occur in organisms that have only perceptual representations.

Dretske considers the possibility of error to be a crucial prerequisite for a satisfactory theory of representation. However, he does not show how cognitive systems can not only make errors but also detect their own error. Thus, he does not address how the representing cognitive system might detect its own error. In the following chapters, I will argue that the ability to detect the adequacy of its own representations is necessary for a cognitive system to have access to the correctness of its own representations.

7.4 An Example of Misrepresentation: Erroneous Misrepresentation in Frogs

Dretske himself denies that a toad has the capacity for mental representation, as it does not learn and does not meet the basic requirement of his concept of mental representations:

> (…) those creatures who have separate detector systems for each nutrient they need and inflexible (not susceptible to change by learning) responses to each, have, on this account, no experience of the food their detectors detect. Since they have no cognitive system for the detectors to service, no way of recalibrating in terms of individual needs and circumstances, their representations of food do not qualify as experiences of food [3: 173].

However, Dretske is mistaken about the facts: there is empirical evidence that amphibians can learn and flexibly change their diet composition [12]. Therefore, his reservations about the learning abilities of frogs can be dismissed, and his concept of misrepresentation can be applied to the example of a frog misrepresenting a bee-bee as a fly.

According to the concept of representation as a function of indicating, the representational system of a frog has the function of indicating food. A properly functioning indication mechanism in the frog's body points to a fly by indicating the characteristic of being edible for the frog. It is the representational system of the frog as a whole that has the function of indicating the presence of food, in part because a component of the frog's system, such as its eye, is equipped with a representational mechanism that allows for the visual detection of food location. For instance: the representational system of the frog's body has the function of registering the presence or absence of food because there is a teleological-informational relationship between the representational state of the frog's brain and the representational state of the optic tectum—a part of the frog's brain. The retina of the eye receives all signals concerning the location and movement of objects within the frog's visual field. Meanwhile, the neuronal states in the optic tectum only receive certain signals from the eye: signals of longitudinal movement of elongated objects (specific configurations of visual stimuli).

The frog has an indication function for pointing out a small, moving object in place m because it serves as its prey. The act of indicating takes place thanks to the fact that the frog's eye transmits visual information about the presence or absence of food on the frog's horizon, and due to the representational mechanism in the frog being geared towards the detection of food in the form of a fly. A specific condition of the retinal cover d, which is a property of the frog's optical cover p, represents the presence of food in place m. According to the definition of indicating (see Sect. 7.3), d triggers a sequence of responses in the frog's body culminating in the tongue being shot out towards the fly G, which means that the fly in m has the property of being edible F.

Let's assume that the frog's optical cover transitions into a state that is similar to when a fly is perceived, but in fact, it is not caused by a fly, but by a bumblebee or a button. In this case, we are not dealing with an indicator (or detection), but with something that plays the role of a proper indicator, without actually being one. This is precisely how the frog misrepresents.

7.5 The Disjunction Problem of in Dretske's Theory

The teleological nature of the function allows Dretske to deal with both the horizontal and vertical alternative problem. The disjunction problem arises when it is not possible to distinguish a change in the extension of a term from its incorrect use. The horizontal problem concerns the distality of the cause of representation, while the vertical problem arises when it is not possible to determine which of the possible causal factors in a given link of the causal chain is the intended referent.

In the cited example of the hunting frog, the horizontal disjunction problem occurs when it is not possible to determine whether the state of the optic tectum is caused by a previously encountered fly or by a bee-bee that appeared some time later. The teleological nature of the indicative function allows for unambiguous reference, as the frog's representational mechanism has the function of indicating its food; therefore, the state of the tectum is a natural indicator of the fly, not the bee-bee.

The vertical disjunction problem arises when it is unclear what the referent of a natural sign (neuronal state) is, that is, whether this state designates (a) a fly in location m, (b) a small, moving object in location m, or (c) frog food in location m.

According to Dretske's teleological-informational conception, the frog's representational mechanism primarily indicates food, thus (c) frog food in location m, and only subsequently (b) a small, moving object in location m [1].[4]

In Dretske's later account [3], which utilizes the historical notion of biological function, the disjunction problem is addressed by the concept of selection *for* a trait, which I discuss in the chapter on Millikan's teleosemantics.

7.6 Summary

Dretske's account of mental representations is the first among those discussed to so strongly emphasize the role of cognitive architecture. This is an undeniable advantage of a systemic approach to mental representations, where the cognitive system is no longer understood in terms of a black box. A consequence of their inclusion in a certain system is the assignment of the function of representation at both the personal and subpersonal levels, provided that indicators from the subpersonal level are connected with the functioning of the entire system (as is the case with the frog).

The fundamental advantage of Dretske's account is the emphasis on the role of misrepresentation. This concept not only meets the desideratum II.3.2 of the possibility of misrepresentation but does so in a naturalistic manner, unlike the semantics of truth conditions. The concept of misrepresentation is explicated in terms of dysfunction in indicating, and the indicating relation is identified with the covariation relation. However, this concept of misrepresentation insufficiently emphasizes the role of the cognitive system. It does not sufficiently highlight the system's role in error detection, which would fulfill the desideratum II.3.3 about the system's ability to detect errors.

The naturalization of intentionality in terms of indication function meets the desideratum II.2 of the naturalization of intentionality, as it allows for a naturalistic explication of the concept of intentionality. Reference and, to some extent, also the content of representations depend to a certain degree on the reality represented, which is external to the cognitive system, and in Dretske's view, indicating is external. Moreover, the function of indicating objectively determines the conditions of correctness, which is significant for realism about representation. Since the indication function implies satisfaction conditions, Dretske's account resolves the hard problem of content. The teleological nature of the function allows Dretske to deal with both the horizontal and vertical disjunction problems.

Desideratum II.3.1 concerning productivity poses a problem for Dretske, as it requires demonstrating that the productivity of parts of representation is present in his conception. Dretske's account of indication provides some form of systematicity, but since it does not guarantee compositionality, it neither ensures nor requires productivity.

[4] According to Dretske, perception indicates distal features of objects, not the proximal features of receptors [1].

Dretske acknowledges the possibility of error as a crucial element of an adequate theory of representation, though he does not show how cognitive systems can detect their own errors. Nevertheless, Dretske's teleosemantics offers a general framework for theories of mental representation, and his solution to the problem of the possibility of misrepresentation can be enriched by the assumptions of other teleosemantic concepts [13–15].

References

1. Dretske, F.I.: Knowledge and the flow of information. Cambridge University Press (1981)
2. Dretske, F.I.: Misrepresentation. In: Bogdan, R. (ed.) Belief: form, content, and function, pp. 17–37. Clarendon Press (1986)
3. Dretske, F.I.: Naturalizing the mind. The MIT Press (1997)
4. Cummins, R., Roth, M.:. Meaning and content in cognitive science. In: Schantz, R. (ed.), Prospects for meaning, pp. 365–382. de Gruyter, Berlin & New York (2012)
5. Cummins, R.: Representations, targets, and attitudes. MIT Press, Cambridge (1996)
6. Ramsey, W.M.: Representation reconsidered. Cambridge University Press, Cambridge (2007). https://doi.org/10.1017/CBO9780511597954
7. Morgan, A.: Representations gone mental. Synthese **191**(2), 213–244 (2014). https://doi.org/10.1007/s11229-013-0328-7
8. Miłkowski, M.: The hard problem of content: solved (long ago). Stud. Log., Gramm. Rhetor. **41**(1), 73–88 (2015a). https://doi.org/10.1515/slgr-2015-0021
9. Dretske, F.I.: Dretske's replies. In: McLaughlin, B. (ed.) Dretske and his critics, pp. 180–219. Cambridge University Press (1991)
10. Dretske, F. I.: Explaining behavior: reasons in a world of causes. Cambridge University Press (1988)
11. Dennett, D.C.: Do-it-yourself-understanding. In: Brainchildren, pp. 59–79. MIT Press, Cambridge (1998)
12. Schmajuk, N.A., Segura, E.T., Reboreda, J.C.: Appetitive conditioning and discriminatory learning in toads. Behav. Neural Biol. **28**(4), 392–397 (1980). https://doi.org/10.1016/S0163-1047(80)91698-2
13. Millikan, R.G.: Language, thought, and other biological categories: new foundations for realism. MIT Press, Cambridge (1984)
14. Millikan, R.G.: Language: a biological model. Oxford University Press, Oxford-New York (2005)
15. Bickhard M.H: The interactivist model. Synthese 166(3), 547–591 (2009)

Chapter 8
Millikan's Account of Misrepresentation

The primary aim of the naturalistic conception of Ruth Garrett Millikan—as well as that of Fred Dretske—is explaining intentionality in terms other than itself, so as to account for misrepresentation. However, Millikan's conception differs in many respects from Dretske's approach, even though they share the etiological conception of proper function (here referring to Dretske's later conception [1]. According to Millikan, what is crucial is what happens with the information received at the output of the representing system.

Millikan addresses misrepresentation in terms of proper function. I will first discuss the concept of proper function, and then analyze the conception of representational error as dysfunction. I will particularly focus on the advantages and disadvantages of this conception of error.

The plan of the chapter is as follows. First, I will present the producer–consumer system describing representation in terms of historical proper function (Sect. 8.1). I will show that Millikan's conception of representation is richer than those previously discussed, partly because its application allows for the description of individual and social behaviors of representational systems (Sect. 8.2). Thanks to the developed model of representation, Millikan can account for the system's misrepresentation (Sect. 8.3).

8.1 Proper Function

In her teleosemantic conception, which she calls biosemantics, Millikan utilizes the etiological concept of proper function. This function is attributed to traits that have survived as a result of continuous reproduction that accounts for the survival of their bearers [2: 116–117]. For instance, possessing a heart is a trait that contributes to the survival of a selected species of frogs.

K. Bielecka, *What is Misrepresentation?*, Studies in Applied Philosophy, Epistemology and Rational Ethics 71, https://doi.org/10.1007/978-3-031-84375-4_8

A feature of function is the duplication of traits within the population of their bearers. However, for a function to earn the title of 'proper', it must arise as a result of selection:

> To have a proper function an item must also come from a lineage that has survived due to a correlation between traits that distinguish it and the effects that are "functions" of these traits, keeping in mind that a correlation is defined by contrasting positive with negative instances. [2: 116]

Unlike [3], who understands function in synchronic terms, Millikan refers to the biological history of organisms (or their parts). The result of a functional analysis in the sense of Cummins is a description of the structure of a given system—often referred to in the literature as design—which entails the decomposition of a system into its constituent parts that causally interact with one another. As John Maynard Smith asserts, "Where an engineer sees design, a biologist sees natural selection" [4: 177]. While Maynard Smith views natural selection through the lens of genetic information transfer, Millikan equally emphasizes the role of the environment and phenotypic factors.

Natural selection acts on a combination of diverse traits that contribute to reproductive success and selectively maintains them [5]. It involves the selection of certain traits in a given environment that contribute to reproductive success (selection for, or selection for a trait). Elliot Sober defines selection for a characteristic as follows:

> Characteristic *c* is an adaptation for doing task *t* in a population if and only if members of the population now have *c* because, ancestrally, there was selection for having *c* and *c* conferred a fitness advantage because it performed task *t*. [6: 85]

The characteristic of having a heart (characteristic *c*) is adaptive for pumping blood (task *t*) in a given population if its members have hearts as a result of previous selection for having a heart and because having a heart contributed to their fitness by pumping blood, not because of its beating, even though the beating of the heart and the pumping of blood co-occur. Millikan describes selection for a trait as follows:

> Intuitively, these traits have been selected for reproduction over actual competitors. Because the correlation must be a result of a causal effect of the trait, the trait will not merely have been "selected" but will have been "selected *for*". [2: 116–117]

Thus, object P has a proper function if it derives from a genealogical line that has survived as a result of the correlation between its distinguishing features and what those features serve (their function). The features of this object were selected as a result of the process of natural selection. For example, the giraffe's neck has (among other things) the proper function of obtaining food from trees because it has survived, among other reasons, because of its length (trait), which facilitates the giraffe's access to food (function realization).

In this context, the criticism posed by Fodor, which suggests that the extensionality of natural selection leads to the indeterminacy of content, seems misplaced. Natural selection is extensional in the following sense: if for organism O it is adaptive to perform action M in the presence of environmental feature F, and F is reliably coextensive with another feature G, then it is equally adaptive for organism O to perform action M in the presence of G [7]. Fodor objects that any environmental feature, correlated with another feature to which the organism has also adapted evolutionarily, triggers the same action in the organism as the first feature. For example, the feature of being a fly correlates with the feature of being black, and both, according to Fodor, cause the frog's reaction of extending its tongue toward the insect. However, Fodor overlooks the existence of selection for a trait, which justifies the choice of one feature over another as functional: according to natural selection, features are ultimately chosen for the survival and reproduction of the species. It is natural selection that indicates and limits the set of possible features for which the function of a part or the whole organism is determined [8]. Therefore, this criticism does not seem particularly devastating. Notably, Fodor's criticism of evolutionary theory has become so pronounced that it has drawn not only criticism but also a degree of hostility from some philosophers of biology, as exemplified in the review written by Block and Kitcher [9].

A proper function may fade and evolve over time, transforming into derived functions or exaptations. The concepts of derived functions and exaptations serve to explain how things "seemingly very new indeed under the sun" [2: 131] can have functions. For example, a derived function from the proper function of matching a chameleon's skin color to the background is the function of changing the skin to green when the chameleon sits on something green. The concept of a derived function allows us to describe the operation of the proper function (e.g., the function of changing color) in a certain environment, such as one where the color green is present. Thanks to derived functions, the relational structure of the proper function is replicated:

> It is easy to suppose that a function is new or that it must be separately described because its adapted or derived character is novel, whereas a closer look would show that under a relational description it is the same old function over again. [2: 125]

An exaptation, on the other hand, is a particular kind of adaptation that arises incidentally and serves a different function, not the one from which it originated.

> Traits already present in an organism for quite different reasons are co-opted for use along with newly fashioned traits to serve new functions. Thus they end up helping to serve functions that are proper functions, but that are not their own proper functions. [2: 140]

For instance, the sea otter uses its fat-covered belly as a table for eating in the sea—a trait that never before served a proper function of this kind, but began to do so under new environmental conditions.

Determining function involves assigning a function to the bearer B of some trait T. Functions are the result of a selective history, during which they were selected for a historically stabilized trait that favors the survival and reproduction of the organism. The bearer of a function is a certain type, and secondarily, an token of that type. For example, the function of indicating a fly can be attributed to both the frog-eye type (where the type is the organ) and the frog-eye token (where the token is a single frog eye, falling under the frog-eye type). Similarly, the proper function of pumping blood can be attributed to both the heart-token and the heart-type. The function of the frog's eye was not selected solely for perceiving flies, but also for detecting other moving objects in the environment, which help the frog better navigate its surroundings. The concept of derived function and exaptation serves to describe the change of function as a result of evolution. Therefore, as Dretske points out, not only does the functioning of an organism depend on the environment, but the functions themselves depend on it.

It is worth examining certain consequences of the etiological concept of function. They are such that the first tokens of a certain type do not have functions, as there is nothing that would be a previous token of that type. Millikan dismisses this objection, noting that it would only concern the very first biological system on Earth. A more challenging objection concerns systems with a causal structure identical to the structure of a system resulting from relevant historical processes, but not replicating the structure of previous tokens. This problem is illustrated by Davidson's thought experiment with the Swampman. The aim of his experiment was to cast doubt on the thesis that intentionality can arise without history. Since intentionality depends on biological function in Millikan's view, this experiment can also be related to her conception. The experiment narrates a fantastical scenario where, as a result of an unexpected lightning strike in a swamp, an ideal molecular replica of Davidson called the Swampman is created. According to Davidson, none of the Swampman's intentional states (beliefs, desires, utterances) have content.

> My replica can't recognize my friends; it can't recognize anything, since it never cognized anything in the first place. It can't know my friends' names (though of course it seems to), it can't remember my house. It can't mean what I do by the word 'house', for example, since the sound 'house' it makes was not learned in a context that would give it the right meaning—or any meaning at all. [10: 443]

Davidson asserts that intentional properties do not arise without history. Millikan fully concurs with Davidson's interpretation of the thought experiment. Fodor succinctly summarizes her position: "it's sufficient to make Swampman's head contentless that he lacks an evolutionary history" [11: 126]. Davidson's idea is utilized by Millikan to defend the etiological concept of function, in which a cognitive system can be ascribed representations, and specifically representational content, only when the mechanisms generating this content have previously existed in its evolutionary history. Thus, the only justification for the intentionality of mental states is history (evolutionary): content cannot be intentional unless there are mechanisms generating representations that stem from its evolutionary predecessors. While Millikan does not deny the possibility of ascribing to the Swampman representations in a metaphorical sense (in terms of Dennett's intentional stance), the attribution of actual representations is unwarranted: the Swampman could not have acquired any experiences in the environment in which he found himself by chance, nor did he have any evolutionary predecessors [12].

However, the thought experiment with the Swampman is treated as an objection to the etiological theory of function [13–15]. Many authors do not share Davidson's intuition, believing that the Swampman could indeed think and learn. The discussion also brings up other examples of replicas of functional systems, for instance, one can imagine the creation of a lion replica that would not have the appropriate evolutionary history, and thus it could not be said that such a lion has a dysfunction of biological organs when it is bitten by another lion.

Millikan might respond to this objection by claiming that the emergence of the Swampman or a miraculously replicated lion is extremely improbable. Living organisms are the product of historical processes and are subject to natural selection. Although the thought experiment with the Swampman indicates certain conceptual limitations of the etiological concept of function, it is not a problem that scientific theories will encounter in research practice, and Millikan's conception is meant to correspond to this practice, not to any intuitions. Richard Dawkins emphasizes that the only available explanation for the existence of highly organized biological systems is an explanation in terms of natural selection:

> If you find something, anywhere in the universe, whose structure is complex and gives the strong appearance of having been designed for a purpose, then that something either is alive, or was once alive, or is an artefact created by something alive. It is fair to include fossils and artefacts, since their discovery on any planet would certainly be taken as evidence for life there [16: 16].

In other words, Millikan does not consider this objection significant, as her conception will be able to address all the interesting cases of real organisms, and she dismisses its inconsistency with intuitions regarding highly improbable counterfactual situations as trivial. Even if this objection is taken seriously, for an account of representation that is primarily intended to correspond to scientific practice, it is not particularly burdensome. This is the cost of a relatively simple definition of function, which largely aligns with scientific practice, but does not fully match all intuitions (such consequences are not present, however, in the hybrid concept of function, which I will discuss in this chapter). Nevertheless, Millikan rules out the possibility of the emergence of a new producer and consumer mechanism that would also be functional. This is precisely the objection that Mark Bickhard and Richard James Campbell raise against Millikan. I will take a closer look at this objection in this chapter, where I discuss Bickhard's synchronic function of representation.

8.2 Proper Function of Representation

An organism (or a part thereof) possesses the proper function of representing F when a particular trait T has been selectively favored throughout the evolutionary history of this organism's lineage due to F. This implies that in accurate explanations for the presence of this trait, it is positively correlated with the fulfillment of function F and leads to the replication or dissemination of trait T due to the effect for which this trait underwent natural selection. The bearer of the proper representational function is the producer–consumer system. The producer subsystem tokens representations, to which the consumer suitably responds (I discuss the concept of producer and consumer further below). The outputs of the producer are particular representational states, which possess a derived proper function of representation. For instance, the frog has the proper function of representing flies when a trait of the frog's visual apparatus has been selectively favored in the evolutionary history of frogs for this function. This means that accurate explanations for the presence of this trait indicate its positive correlation with the performance of the function; this leads to the further replication or dissemination of the trait precisely because of satisfying the frog's hunger (the effect), and thus the trait was favored by natural selection. The bearer of the proper representational function in frogs is the producer of representation, namely, the visual system, and the consumer of representation, that is, the motor system. Thus, the state of fly representation is an output of the producer, to which the consumer appropriately responds. The producer's outputs possess a derived proper function, that is, the specific token of the frog's representation of a fly.

In Millikan's conception of functional explanation, she refers to the so-called Normal explanations.[1] These concern the manner in which a historically stabilized lineage of certain organisms or their parts fulfills a given proper function. For the realization of a given proper function, traits that organisms possess are necessary, thanks in part to natural regularities and the fact that they have been selected for that function [17]. Essential components of a Normal explanation, the so-called proximate Normal explanation, include the so-called Normal conditions. For example, the Normal conditions for the proper functioning of the heart encompass its external and internal structure (Normal properties and structures), as well as regular dependencies with other parts of the organism and environmental features that influence the heart's correct operation (the regularity of electrical impulses sent to the heart, the heart's oxygen supply, and the function of the closed circulatory system that provides the heart with oxygen) [17]. Millikan emphasizes that Normal explanations primarily concern the regularity of functioning, while the origin of functions is of secondary importance. This has consequences for the proper function of representation, for which the Normal explanation primarily concerns the semantic rules of mapping.

It should be stressed that Millikan's intention was to clearly distinguish the concept of Normal conditions from the concept of normal conditions previously used in the literature. She thus rejects the conception of conditions for the proper operation of functions, adopted among others by Gilbert Harman [18] and John Bigelow and Robert Pargetter [19]. Unlike these authors, according to Millikan, it is not the frequency of occurrence of certain traits that is decisive for function, but rather that at least once a trait has been used in an evolutionarily advantageous way, that is, it has been selected for that trait. Millikan therefore rejects the frequency of occurrence of a trait as a criterion for distinguishing a functional trait. She justifies this by citing biological examples:

> The business of the biological species of staying in business determines standards for individuals of that species, standards which, though they often correspond to averages, are not defined in terms of mere averages over the species. Consider, for example, how few sperm

[1] Millikan uses the term "Normal" with a capital 'N' to denote the explanation of essential—not merely typical—factors that determine the fulfillment of a function.

> or immature members of most species actually manage to perform all the functions that nonetheless are proper to them. [17: 4]

Biological functions need not be fulfilled frequently, and evolutionary conditions often diverge from the statistical norm. An example of this can be seen in certain phenotypic traits of plants in the Atacama Desert, which have been naturally selected to perform their function when it rains, even though this occurs only once every hundred years.

Normal conditions do not constitute a definitive list of factors determining the emergence of the proper function of representation. On the contrary, Millikan considers this list to be open-ended, thereby countering the potential criticism that Normal conditions must be presupposed as given in order to explain how a function is realized. Instead, Normal conditions are not given as in pre-representational harmony but are formed throughout evolutionary history.

The most immediate Normal explanation elucidates not only how a given function may be realized but also its origin. However, the role of a Normal explanation is not to provide all factors related to the origin of the function, but only the most crucial ones. The most immediate Normal explanation of representation primarily clarifies the emergence of specific semantic mapping rules, that is, correspondence rules between signs and states of affairs that were established in the past when the consumer and producer (or their predecessors) performed this function. Only when mapping rules are established can we speak of individual mappings that fall under these rules. The bee dance serves as an example. However, I will first discuss the producer–consumer system. It is important to emphasize that, contrary to many interpreters of Millikan [7, 20–22], mapping rules (in particular, the relevant mappings, not the mapping relation itself) are explained (constitute the explanandum).

A specific representational mechanism acts as the bearer of mappings of a given type. As changes occur in the history of the representational mechanism, the corresponding types of mappings evolve as well. Types of mappings are contextually dependent on time and place and can also form a network of connections with other mappings, including those produced by other cognitive systems. Mappings form a structure of signs connected by relations (they are compositional[2]).

Millikan introduces the concept of "locally recurrent natural signs" to describe the components of representation (or intentional signs[3]). Local signs are "recurrent," meaning they are tokens of the same types of signs only when they are part of a larger whole, that is, when a sign token belongs to a contextualized possible class of some type:

> … what "recurs" in the case of recurrent signs are other members of the same *system* of signs, where a sign system consists in a set of possible sign types, designatable transformations of which ("transformations" in the mathematical sense) correspond systematically to transformations of what they signify. [12: 49]

---

[2] Millikan emphasizes that this is not linguistic compositionality, but a relational structure [12].

[3] The term "intentional signs" is Millikan's designation for representations.

In other words: individual tokens of mappings of a given type arise as a result of transformations (which provides a rather minimally understood productivity—minimal because the transformations may be limited to a fairly small set of possibilities), and these transformations, in turn, systematically correspond with the transformations of what they signify [12]. The proper function of the representational mechanism is to ensure that local natural signs reappear in the presence of the appropriate objects or features. Central to Millikan's conception is that mapping rules are semantic rules designed for the organism, as they are meant to serve its orientation in the world. In other words: mappings are meant to guide its behaviors in the environment:

> Millikan claims that a true representation maps onto its represented *in accordance with semantic rules* [emphasis - K.B.] determined by the way the systems using the representation are designed to react to it in guiding, perhaps first inference processes, but ultimately behavior. [23]

The function of representation is attributed to particular mechanisms due to their nearest Normal explanation. Like Dretske, Millikan describes the representation mechanism as an integral part of the cognitive system. It is crucial to note that Millikan understands the representational mechanism more broadly, recognizing that there are parts of the cognitive system that utilize representations. She believes that the organism's structure and the mechanisms it contains condition the ability to represent. Thus, she develops the idea present in Dretske's work that the organism's structure and its functional organization significantly influence representation. Millikan reduces the ability to represent to the sub-abilities of the mechanisms that comprise this ability. Like Dretske, she considers content understood in informational-causal terms as key to determining the ability to represent.

Among the mechanisms necessary for the function of representation, Millikan distinguishes the tracking mechanism.[4] This is the minimal ability of the representational system that forms the basis for reidentification, that is, the re-identification of features or objects of a given type found in the organism's environment. The ability to track is one of the fundamental representational abilities. The proper function of the tracking ability is to produce temporary internal signs that assist in representation, enabling the accurate mapping of reality [17: 173].

Tyler Burge, like Millikan, emphasizes the role of tracking mechanisms in perception: he considers the tracking mechanism to be constitutive of perception and present at a very basic level [25], he also highlights the role of perceptual constancy mechanisms. Perceptual constancies are fundamental to an organism's ability to detect objects and features in the world around it. I describe perceptual constancy mechanisms in more detail in Chap. 10, where I argue that they seem insufficient to define the concept of representation in a non-trivial way. Certainly, however, tracking abilities and the abilities to detect perceptual constancies allow us to speak of the organism's capacities to detect and distinguish many properties; they do not, however, allow for

[4] It is worth noting that the literature also uses the term 'tracking' to describe informational relations such as indicating (see [24]). However, that is not the sense of the term at issue here.

the recognition of inconsistencies, which I consider a touchstone of the ability to represent.

The complex ability to track consists of the organism's ability to recognize the location of an object or its feature, as well as the ability to continue tracking other locations of the object or feature. These abilities also serve the reidentification of the object. Directing the gaze elsewhere, for example, to the right, facilitates the identification of another instance of the object (or feature) of the same type. I will illustrate this with the example of a frog looking at a fly. The frog's eye has the ability to track the fly, provided there are detectable properties of the fly (such as its shape and color). The frog has the ability to track the fly at location l_1 at time t_1, and to track the same fly at location l_2 at time t_2. Since its eye does not perform saccadic movements, the frog must turn its entire head to follow the insect. This may allow it to reidentify the same instance of the fly.

It is worth noting that Millikan, unlike Dretske, does not separate the sensorimotor apparatus from the mechanism responsible for its conceptual abilities.[5] In Millikan's view, innate learning mechanisms for identifying (and reidentifying) certain types of objects (and features) are responsible for the conceptual grasp of reality. For instance, certain mechanisms in the frog have been selected for recognizing individual species of animals. More developed organisms, such as humans, have more complex mechanisms of identification and reidentification, particularly cognitive ones.

Millikan presents representational functioning in the form of a producer–consumer system. The function of representation is assigned to a representational mechanism composed of a system called the "producer" and a system called the "consumer," between which representations are transmitted. The producer and consumer mechanism arose as a result of natural selection or through learning. Millikan drew inspiration for creating this system from the linguistic sender-receiver model.

Thus, Millikan conceives the function of representation relationally, as composed of the relations between functions assigned to the producer and consumer remaining in evolutionarily stabilized cooperation:

> Representation consumers are devices that have been designed (in the first instances, at least) by a selection process to cooperate with a certain representation producer. The producer likewise has been designed to match the consumer. [38: 126]

The role of the producer is not to generate any representations, but those whose form reflects relations with the world. Therefore, the role of the producer is the production of mappings.

> The representation producer has been designed through selection to generate representations for the consumer that correspond to conditions in the world by virtue of a correspondence rule present in the nearest explanation of the consumer's normal successes. Put simply, the producer's role is not only to create representations—graphically, the "shape"—but to produce mappings, a certain relationship between the "shape" and the world. [38: 127]

[5] In [26], Millikan rejects the existence of socially shared concepts in favor of unicepts. A unicept is a psychological equivalent of a concept. It defines the identity of things formed through various stimuli. A unicept determines the common meaning of learning, reasoning, and directing action.

Importantly, and often overlooked by many interpreters of Millikan [7, 20, 22], mappings exist not only thanks to the producer but also as a result of the interaction between the producer mechanism and the consumer mechanism. Producers cannot create representations that remain in arbitrary relations with the world, but only in correspondence with it, which is a condition for the proper functioning of consumers:

> Its [the producer's - K.B.] is to produce representations, "shapes," that *correspond*, that is, correspond in the manner required by the representations' consumers in order that these consumers should function properly. [38: 130]

Thus, both the producer and the consumer play an equally important role in the creation of these mappings. The proper function of the representation producer mechanism is to map such (objective) properties whose presence can be detected by the consumer mechanism. The role of the consumer, therefore, is to receive the contents delivered by the producer, because the consumer has been selected for the reception of information about properties or states of affairs transmitted by the producer.

The cooperating mechanisms of the producer and the consumer constitute the mechanism for generating representations of a given type. The representational mechanism appears and evolves in the evolutionary history of organisms of a given type. The functioning of the representational mechanism depends on the structure, environmental conditions, including other organisms (which encompasses the concept of Normal conditions). Millikan emphasizes that the stimuli appearing at the input (producer) are not the primary condition for the content of representations; they are secondary to what happens next (consumer). Nevertheless, the consumer is sensitive only to a certain repertoire of signs (belonging to specific types). It represents not proximal features, such as color or shape, but distal features, such as being edible. It is because of distal features that an organism adapts to the environment. This consumer-oriented teleosemantics of Millikan differs from the producer-oriented teleosemantics defended by Karen [7, 20, 27].

It's important to note that the common interpretation of Millikan's teleosemantics insufficiently emphasizes the role of the consumer in determining mappings [7, 20, 28, 29]. In the common interpretation, the producer–consumer system is based on a correlational concept of content, therefore correspondences must be pre-given [28–31]. Hutto and Myin further accuse Millikan of not solving the hard problem of content, as they believe a correlational theory of content is too weak for this purpose. I have refuted this objection in Chap. 7, using Dretske's conception as an example, showing that Dretske had already solved the hard problem of content.

[17] illustrates the role of mappings in the producer–consumer representational system with the example of the bee dance, originating from Karl [32]. The producers of representations are the bees performing the dance and representing the location of nectar, while the consumers are the bees observing the dance of others and receiving the signal about the location of the nectar. The mapping rule is thus a general regularity of the bee's movement from the location of the nectar, and the mapping itself is a specific dance in relation to the chosen nectar.

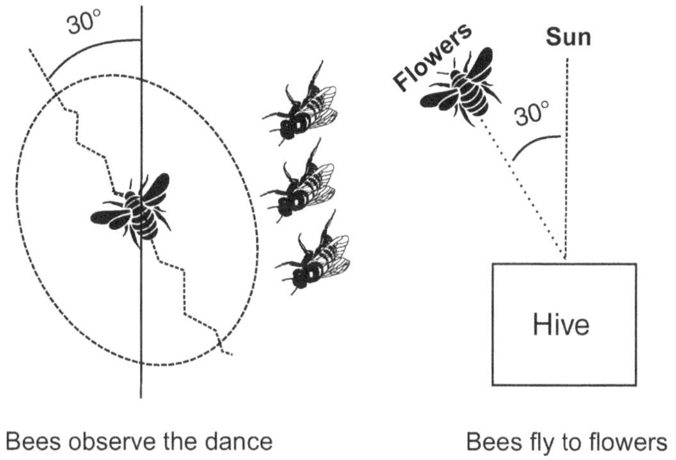

Fig. 8.1 Behavioral patterns of bees

Frisch observed that bees, upon finding nectar at least about 100 m away from the hive, return to perform a specific kind of dance. They dance tilted towards the vertical at an angle that reflects the inclination of the food relative to the azimuth of the sun. The angle of the bee's movement currently towards the vertical reflects the angle of the food's location relative to the current azimuth of the sun. For example, if the food (nectar) is located in a straight line (relative to the azimuth of the sun), which is to the north, the bee's dance is directed straight towards the vertical; however, if the food is located 30º to the west of the hive, the dance is oriented to the right of the vertical (see: Fig. 8.1). The duration of the dance and the length of the path traveled during the dance, as well as the buzzing produced by the wings, map the distance of the bees to the location of the nectar [33].

The bee dance causes other bees observing it to find the nectar. Therefore, the dance maps, according to the Normal explanation, the actual position of the nectar, the sun, and the hive.

The producer–consumer system serves to describe various representational systems. Although this system naturally describes the relationships between organisms, it equally applies to relationships between systems within a single organism. To describe representation in simple organisms, Millikan [34] uses the model of so-called pushmi-pullyu representations. They are descriptive-directive, meaning that a representation is always not only a description but also an action instruction.

For example, a frog representing a fly has a representation in the form of an instruction transmitted by the producer (the moving black dot), which is received by the consumer, for whom the content of the representation, stimulating the frog to flick its tongue, is "eat now here."

The producer–consumer system as an originally communicative model easily captures linguistic behaviors as representational actions. However, they are not specially distinguished—they merely constitute a subset of other representational behaviors. The communicative model of producer–consumer is consistent with the evolutionary approach to language, according to which it arose as a result of transformations of simpler communication systems.[6] For similar reasons, Millikan considers social behaviors (e.g., of bees) as one of many representational behaviors, although she recognizes individual representational behaviors as equally important. An illustration of social representational behaviors in animals is the bee dance; I will illustrate individual abilities to represent using the example of a frog. The discussed theory of representation is more general than Dretske's account, which does not accommodate social behaviors as representational.

However, despite several undeniable advantages, Millikan's model has undesirable consequences for the conception of representation. Too many processes qualify as representational, making the concept too broad. This model does not exclude tropisms and taxes, which prima facie do not require a description in representational terms. Burge shows that non-representational interactions of organisms with the world are entirely sufficient to explain them [25]. He defends the thesis that at the base of such behaviors lie not representations, but sensory registrations:

> Such sensory registration, even together with the quantitative transformations, is not perception. It tends to correlate well with environmental conditions. It functions to enable the individual to respond to environmental conditions in a fine-grained way. But it can be fully explained in its *operation* without appeal to objectification.[7] [25: 398]

Burge accuses Millikan (and other proponents of teleosemantics) of considering registrations as representations. He asserts that there is no need to adjudicate on the objects of registration, whereas representations have objects. Millikan's explanation of tropisms in evolutionary and adaptive terms would also be representational, even though the cognitive abilities required in these cases are too limited to attribute the capacity for representing to the organism. In this sense, Millikan's conception does not meet Ramsey's job description challenge, as her concept of representation is prone to trivialization.

Importantly, the ability to register does not require the ability to misrepresent; it is difficult to speak of error-making by organisms capable of, for example, tropisms, since tropism is not at all sensitive to error. Thus, Millikan considers as representational even those abilities that do not require the capacity for misrepresentation. However, if this is the case, the possibility of misrepresenting ceases to be a test of mental representation.

---

[6] The relationship between representation and language is a very broad topic [35], which goes beyond the scope of this book. However, it is worth adding that due to externalism regarding content, the role of communication, and the biological nature of Millikan's conception of representation, it is close to biosemiotic explanations of the evolution of meaning as a result of the organism's interaction with their environment [36].

[7] By "objectification," Burge understands the possession of an object by representations that relate to distal features and objects, rather than to the features of sensory stimuli.

8.3 Misrepresentation as Dysfunction

Millikan emphasizes the role of teleology as primary to the establishment of mapping rules in explaining the possibility of error-making by an organism [37]. Dysfunction consists in the failure to fulfill the proper function. For example, cardiac dysfunction is the failure to fulfill the heart's proper function of pumping blood throughout the body.

We can speak of dysfunction when certain conditions from the series of Normal conditions of a given proper function are not present, or more precisely: only when individual instances of the type that has this function are not in certain Normal conditions. For example, cardiac dysfunction occurs when the Normal condition of having a sufficient amount of oxygen in the environment in which the organism resides is not met, resulting in the heart not being sufficiently oxygenated and not functioning properly.

Representational error is conceptualized in terms of the dysfunction of representing. For instance, a frog has a dysfunction in representing, say, flies when, under certain conditions (the presence of bumblebees), one of its producer mechanisms (its visual apparatus) or consumer mechanisms (its motor apparatus) does not function correctly, leading the frog to perform an action inadequate to the represented object [38].

Error or representational dysfunction can often occur. It is worth repeating that, according to Millikan, the frequency of dysfunction occurrence is irrelevant. Millikan illustrates this with the example of beavers, which slap the water surface with their tails whenever they hear disturbing sounds that usually signify an approaching predator. Beavers often erroneously represent these sounds as indicative of danger, but without this excessive vigilance, they could easily become extinct. The occurrence of error is thus justified by evolutionary economy—from an evolutionary perspective, it is more beneficial to excessively, and thus erroneously, represent certain environmental features than not to represent them at all. Therefore, in Millikan's conception, what is essential is the cognitive system's ability to make errors, which the consumer is capable of responding to. This creates, especially considering the consumer mechanism, a conceptual framework for the organism's ability to detect an error by the system. These are only frameworks, as in Millikan's conception itself, the representational error is not detectable without knowledge of the organism's evolutionary history. However, it is difficult to attribute such knowledge to an organism, for example, a beaver. Thus, Millikan does not adequately analyze the concept of an error detectable by the system and does not meet the desideratum about error II.3.3.

8.4 Summary

Misrepresentation is an element of normal functioning in the discussed theory. A certain degree of error in representation is necessary, among other things, due to the limitations of organisms stemming from their structure and organization. As a result, Millikan's justification of the importance of error fits well into biology. It is hard to imagine that in nature, there are organisms that never make errors.[8]

The producer–consumer system allows for a fairly detailed description of both social representations and individual ones. However, due to too liberal a definition of the representing function, Millikan's conception does not meet Ramsey's job description challenge and allows for the trivialization of the concept of representation (but see [40] for a defense of liberality in teleosemantics). It will be attributed to very simple organisms that are only capable of sensory registration, as well as those that are not capable of making errors.

Millikan also does not consider the question of how it is possible for fallible living organisms to detect errors. Since error is a dysfunction, and dysfunction is defined solely in historical terms, there are doubts as to whether organisms can have access to erroneous representations at all (unless they are adaptive biologists). After all, beavers do not know the evolutionary history of their species.

The architecture of the representational system is more precisely described by Millikan than by Dretske. Millikan does not develop this thought further, but within the concept of proper function, there is room for innate fixed abilities of the organism, independent of phylogenetic individual learning. Such abilities may constitute part of the mechanism of representation as well as the mechanism of misrepresentation (more on this in Chap. 10). Therefore, the discussed theory of representation meets the desideratum of the possibility of misrepresentation (II.3.2.), but does not fulfill the desideratum of the cognitive system's ability to recognize the adequacy of representations (II.3.3).

References

1. Dretske, F.I.: Naturalizing the mind. The MIT Press (1997)
2. Millikan, R.G.: Biofunctions: Two Paradigms. In: Ariew, A., Cummins, R., Perlman, M. (eds.) Functions: New essays in the philosophy of psychology and biology, pp. 113–143. Oxford University Press, USA, New York (2002)
3. Cummins, R.: Functional analysis. J. Philos. **72**(20), 741–765 (1975). https://doi.org/10.1007/978-1-4614-7004-5
4. Maynard Smith, J.: The Concept of information in biology. Philos. Sci. **67**, 177–194 (2000)
5. Darwin, C.: On the origin of species by means of natural selection, or the preservation of favoured races in the struggle for life. John Murray, London (1859)
6. Sober, E.: Philosophy of biology. Westview Press, Boulder-San Francisco (1993)

[8] In contemporary research on perception, some argue, using game-theoretic simulations, that veridical perception is evolutionarily disadvantageous [39]. These arguments may go too far, though.

7. Neander, K.: Teleological theories of mental content. In Zalta, E.N. (ed.) The Stanford Encyclopedia of Philosophy (2012). Retrieved from http://plato.stanford.edu/archives/spr2012/entries/content-teleological/.
8. Sober, E.: Evolution and optimality—feathers, bowling balls, and the thesis of adaptationism. Philos. Exch. **26**(1), 40–57 (1996)
9. Block, N., Kitcher, P.: Misunderstanding darwin. boston review. (2010) Retrieved from http://bostonreview.net/ned-block-philip-kitcher-misunderstanding-darwin-natural-selection
10. Davidson, D.: Knowing One's Own Mind. Proceedings and Addresses of the American Philosophical Association **60**, 441–458 (1987)
11. Fodor, J.A.: The elm and the expert: mentalese and its semantics. MIT Press, Cambridge, MA (1995)
12. Millikan, R.G.: Varieties of meaning: the 2002 Jean Nicod lectures. MIT Press, Cambridge (2004)
13. Bickhard, M.H.: The interactivist model. Synthese **166**(3), 547–591 (2009). https://doi.org/10.1007/s11229-008-9375-x
14. Bickhard, M.H., Christensen, W.: The Process dynamics of normative Function. Monist **85**(1), 3–28 (2002)
15. Krohs, U.: Functions as based on a concept of general design. Synthese **166**(1), 69–89 (2007). https://doi.org/10.1007/s11229-007-9258-6
16. Dawkins, R.: Universal Darwinism. In: Hull, D.L., Ruse, M. (eds.) The philosophy of biology, pp. 15–37. Oxford University Press, New York (1998)
17. Millikan, R.G.: Language, thought, and other biological categories: new foundations for realism. MIT Press, Cambridge (1984)
18. Harman, G.: Thought. Princeton University Press, Princeton, NJ (1973)
19. Bigelow, J., Pargetter, R.: Functions. J. Philos. **84**(4), 181–196 (1987). https://doi.org/10.2307/2027157
20. Neander, K.: Misrepresenting & malfunctioning. Philos. Stud. **79**(2), 109–141 (1995)
21. Ramsey, W.M.: Representation reconsidered. Cambridge University Press, Cambridge (2007). https://doi.org/10.1017/CBO9780511597954
22. Shea, N.: On Millikan. Wadsworth publishing, Belmont (2005)
23. Millikan, R.G.: Mental content, teleological theories. In: Chichester, L.N. (ed.) The Encyclopedia of Cognitive Science. Wiley (2006) https://doi.org/10.1002/0470018860.s00128
24. Mendelovici, A.: Reliable misrepresentation and tracking theories of mental representation. Philos. Stud. **165**(2), 421–443 (2012). https://doi.org/10.1007/s11098-012-9966-8
25. Burge, T.: Origins of objectivity. Oxford (2010).
26. Millikan, R.G.: Beyond concepts: unicepts, language and natural information. Oxford University Press, New York (2017)
27. Neander, K.: A mark of the mental: in defense of informational teleosemantics. MIT Press, Cambridge, MA (2017)
28. Hutto, D.D., Myin, E.: Radicalizing enactivism: basic minds without content. MIT Press, Cambridge (2013)
29. Hutto, D.D., Myin, E.: Exorcising action oriented representations: ridding cognitive science of its Nazgul. Adapt. Behav. **21**(3), 142–150 (2013). https://doi.org/10.1177/1059712313482684
30. Cummins, R.: Truth and meaning. In: Meaning and truth: investigations in philosophical semantics, pp. 175–197. Oxford University Press, New York (2002)
31. Shea, N.: Naturalising representational content. Philos. Compass **8**(5), 496–509 (2013). https://doi.org/10.1111/phc3.12033
32. Frisch, K.: The dance language and orientation of bees. Harvard University Press, Cambridge, MA (1967)
33. Shettleworth, S.J.: Cognition, evolution, and behavior. Oxford University Press, Oxford-New York (2010)
34. Millikan, R.G.: Pushmi-pullyu representations. Philos. Perspect. **9**, 185 (1995). https://doi.org/10.2307/2214217

35. Gut, A.: O relacji między myślą a językiem: studium krytyczne stanowisk utożsamiających myśl z językiem. Wydawnictwo UMCS, Lublin (2009)
36. Favareau, D.: Essential readings in biosemiotics. Springer, Dordrecht (2009). https://doi.org/10.1007/978-1-4020-9650-1
37. Millikan, R.G.: Reply to Shea. In: Ryder, D., Kingsbury, J., Wiliford, K. (eds.) Millikan and her critics, pp. 81–86. Wiley-Blackwell, Malden (2013)
38. Schulte, P.: How frogs see the world: putting millikan's teleosemantics to the test. Philosophia **40**(3), 483–496 (2012). https://doi.org/10.1007/s11406-011-9358-x
39. Hoffman, D., Singh, M., Prakash, C.: The interface theory of perception. Psychon. Bull. Rev. **22**(6), 1480–1506 (2015)
40. Artiga, M.: Liberal representationalism: a deflationist defense. Dialectica **70**(3), 407–430 (2016). https://doi.org/10.1111/1746-8361.12152

Chapter 9
Teleosemantics and System-Detectable Error: Bickhard's Interactivist Model

In this chapter, I discuss the conception of system-detectable error, defended by Mark Bickhard in the interactivist model of representation [1, 2].[1] I begin with a review of existing concepts of system-detectable error (Sect. 9.1), then reconstruct the argument for an individually conceived biological model of representation (Sect. 9.2), and discuss the anticipatory conception of representation (Sect. 9.3). I analyze the minimal interactivist model of representation (Sect. 9.4), followed by the comprehensive model (Sect. 9.5). I particularly demonstrate how the discussed conception fulfills the key desideratum of error detection by the system, and consider whether it can be called internalist (Sect. 9.6).

9.1 The Idea of Error for the System in Previously Discussed Concepts

The detectability of representational adequacy by the cognitive system is a necessary condition for representation, as highlighted by Fodor [4], Davidson ([5, 6]), and especially Bickhard [1, 2].

Gilbert Harman was the first to point out that normal conditions should be taken into account in the concept of erroneous representation:

> We should assume at first that circumstances are normal and that no mistakes are made. Then mistakes can be understood as abnormalities. [7: 62]

[1] The idea of the system-detectable error is also emphasized by Michael L. Anderson and Gregg Rosenberg in their guidance theory of representation [29, 30]. Due to the closeness of the approaches of Anderson and Rosenberg and Bickhard, I will limit myself to discussing the latter proposal. See also [3].

K. Bielecka, *What is Misrepresentation?*, Studies in Applied Philosophy, Epistemology and Rational Ethics 71, https://doi.org/10.1007/978-3-031-84375-4_9

Normal conditions allow us to determine what constitutes an erroneous representation. Thus, the normal operating conditions of a radar are the environmental circumstances and those features of the radar itself that are the necessary and sufficient condition for the correctness of representation. They are essential for determining what the radar represents and when it does so. According to Harman, the radar represents if it detects when and under what circumstances airplanes appear, which is recognized by its ability to correctly respond to approaching aircraft (locating them). Since it is known when the radar functions correctly, it can be determined when its readings are erroneous. Therefore, the possibility of correct responding is a necessary condition for system-detectable representation.

Postulating normal conditions, however, poses serious problems for the treatment of misrepresentation. This was most emphatically pointed out by Fodor [4], who noted that normal conditions are used to solve the disjunction problem. This problem arises in correlational or causal theories of representation due to the continual expansion of the set of referents with each erroneous use of a representation. In other words, by using the disjunction operator, another referent is added to the denotation with which a given representation co-occurs. According to Fodor, attempting to frame error in terms of normal conditions is not a good solution, as it relies on the peculiar assumption that there are ideal situations (type 1) and other, non-ideal situations (type 2). However, since there is no operational criterion to distinguish ideal conditions from the rest, it is unclear how to differentiate erroneous representations from correct ones. Moreover, the concept of normal conditions is difficult to operationalize without introducing a teleological function. Unlike the simplest causal theory of mental representation, in teleosemantics, normal conditions depend on biological factors. Their determinant is thus adaptiveness or self-preservation (depending on the concept of function), which are attributed to features of biological systems—the same feature may be functional in one respect but not in another. For instance, the feature of vision that enables humans to quickly detect motion can be dysfunctional, as it leads to the creation of motion illusions.

In teleosemantics, one does not distinguish absolute type 1 situations from type 2 situations; instead, one speaks of the normal conditions of individual functions. However, the concept of normal conditions does not allow for determining whether a cognitive system can detect its own errors, which is (as Fodor, Davidson, and Bickhard indicate) a condition for attributing correct representations—not only does the attribution of misrepresentations depend on whether the system can represent correctly, but the attribution of representations of a given type depends on whether the system can err, thus having the capacity for misrepresentation. An organism cannot have a type of representation that it could never correct (in light of nomological dependencies):

> An organism can't have a kind of symbol which it necessarily misapplies, i.e., which it misapplies in every world compatible with its psychology. (...) There are, after all, some mistakes that can't be made; viz., mistakes from which it is nomologically impossible to recover, consonant with the character of one's psychology. [4: 108]

Magnetotactic bacteria, for instance, cannot have a representation of a composition such as "The Song of the Nibelungs." A biologist might observe that a certain bacterium, in the presence of sound waves caused by a loud rendition of "The Song of the Nibelungs," reacts exactly as if the anthem of the European Union were being played (while other bacteria from the same strain react differently). However, it cannot be said that the bacterium confuses these pieces, because there is nothing in its structure that could help it correct this error, even if this strain of bacteria normally responds differently to these two acoustic stimuli. Therefore, the ability to correct one's own errors is a necessary condition for the representational capacity of a cognitive system.

Davidson independently arrived at the idea of error detectability by a system [5]:

So far as I can see, no account of error that depends on the classifications we find most natural, and counts what deviates from such as error, will get at the essence of error, which is that the creature itself *must be able to recognize error* [emphasis added—K.B.] [6: 140].

Unlike Fodor, Davidson maintained skepticism towards causal theories of representation. His solution is based on the concept of triangulation and radical interpretation, in which intersubjectivity plays a crucial role. The criterion for detectability of representational error, in his view, is not reduced to the interaction of a single cognitive system with the environment, but to the interaction with other cognitive systems.

Davidson's remarks remain very brief. Therefore, one may doubt whether his concept of intersubjective meaning also encompasses situations in which the cognitive system itself detects its own errors. This is quite typical for perception. Intersubjectivity is not needed when the cognitive system can detect discrepancies between what it sees, what it touches, and thus realize that it is mistaken. Davidson's sketchy approach does not suggest a proper development of the idea of the system-detectable error. However, Davidson's intention was to describe the meaning in communication between rational agents.

The issue of detectability of representational adequacy has been cursorily addressed by the concepts of mental representation discussed so far. It is necessary to more precisely capture the role of the cognitive system in detecting representational error. In outline, it is as follows: the cognitive system recognizes certain features or objects in the world, which it then utilizes in its interactions and tests against the world. Bickhard's proposal is the only one that clearly explicates this, thereby fulfilling the second desideratum concerning the system-detectable error.

Bickhard believes that all conceptions that do not allow for the system-detectability of error are flawed:

> One central criterion [for the adequacy of the concept of mental representation - K.B.] is that a model of representation must be able to account for system detectable error: again, if not, then it cannot account for error guided behavior and learning, and is thereby refuted. [2: 568]

It is worth noting that, according to Bickhard, the system-detectability of error is crucial for representation in general. It is necessary to describe the organism's ability to learn from errors. Bickhard elevated the system-detectability of error to the status of a key requirement for an adequate naturalistic concept of representation:

> One requirement to be able to detect error in one's own representations is that the content be accessible, but, with nothing but externally related content available in the models, there is no way in which an organism can access its own content in order to even begin to assess that content in the current environment. [2: 564]

Internal relations are supposed to occur between the bearer of representation and its content (similar to the arc of a circle to the center of the circle):

> An internal relation is internal in a sense of being necessary in order for one or more of the relata to be what it is. (...) One example would be an arc of a circle, which cannot be that arc without bearing a specific relation to the point that is the center of the circle. [2: 562]

When the bearers of representation are cognitively accessible to the organism, potentially so is the content of representation.

9.2 Bickhard's Conception of Mental Representation

The author starts from an assumption typical of systems biology, according to which life is a process occurring in systems that are relatively self-sustaining and require metabolic processes for their existence.[2] Bickhard's interactivism shares assumptions convergent with enactivism in its classic version, defended by Humberto Maturana and Francisco Varela [8–10]. Enactivism posits that cognitive activity is a process resulting from the interaction of an autonomous system with its environment. Maturana and Varela understand this autonomy in terms of autopoiesis. In the autopoietic model, life is the self-organization of the system—a living system is one that maintains its functioning despite being subjected to continuous disturbances. Such a cognitive system is relatively isolated from the environment, has only energetic relations with the surroundings, and co-creates itself. This model *prima facie* excludes death, aging, as well as informational relations with other autonomous systems.

In contrast to Maturana and Varela, Bickhard believes that autonomy should be conceived in physical terms, particularly thermodynamic terms: living organisms remain in the far-from the-thermodynamic-equilibrium state with their environment. In this way, Bickhard, distinguishing himself among other theorists with similar views, such as Hanne De Jaegher [11] and Ezequiel Di Paolo [12],[3] naturalistically explicates the idea of relative autonomy.

---

[2] The interactive model of representation is part of a larger philosophical project in which the approach to the function of representation relies on strong metaphysical assumptions of process ontology. Thus, actions are processes at the level of the entire organism, which break down into individual subprocesses, cooperating within a single integrated cognitive system. Due to the specificity of the subject matter of this work, I will omit a critical analysis of Bickhard's ontological assumptions. However, it is worth noting that process ontology facilitates thinking about representations in interactive terms.

[3] Hanne De Jaegher and Ezequiel Di Paolo describe the relationship with the environment as sense-making, yet they do not explicate what this sense-making is, especially in physical terms.

Bickhard illustrates this with the example of a candle. A candle is capable of burning, and by burning, it sustains its continued burning, melting the wax. The burning wick causes the melting of the wax, which can then burn, and the convection created by the burning wick causes the melted wax in gaseous form to burn close to the wick. The candle is devoid of the ability to choose possible actions: when the wax runs out, its continued burning is threatened, and the candle goes out. Therefore, the autonomy of the candle is very limited.

The assumption that an organism must be active to exist, rooted deeply in American pragmatism, is the fundamental assumption of Bickhard's conception of representation. Without life-sustaining activity, the organism ceases to exist. Bickhard bases his biological conception of representation on the uncontroversial assumption that cognitive systems are biological entities threatened by death. They are characterized by two biological properties that are in mutual tension: mortality and autonomy. The organism strives for autonomy, ensuring its continued existence.

Bickhard's position, akin to that of Millikan, can be most succinctly described as biological pragmatism. Bickhard emphasizes that the fundamental activity of living organisms is ensuring continued survival. Representation is based on the ability to choose between different actions, through which cognitive systems ensure continued existence, and their primary role is to direct actions. The ability to detect the adequacy of representations aids in precise action guidance and increases the chances of success; thanks to the predator's ability to correctly assess the situation, it can capture prey, thereby avoiding hunger. Only a system capable of detecting the adequacy of representations can correct its own errors: a predator has the ability to correct its own actions because it can assess the situation it is in, based on whether the outcomes of its actions deviate from the intended ones. Only then can the predator correct, for example, becoming distracted in the presence of tempting prey, should the prey escape from right under its nose. The ability to correct errors is the essence of representation, by which a cognitive system adjusts its behavior to the environment. Cognitive systems have innate mechanisms for detecting errors in a certain environment, or they begin to better detect their errors through learning. It is worth noting that—unlike Dretske—learning is not a necessary condition for representation.

Living organisms incapable of correcting their own representations may perish. The possibilities for action are not fixed once and for all. Environmental conditions and the representational resources of the system change during its operation. A system that does not respond to these changes is not very flexible and may die.

Bickhard distinguishes between representation in the minimal sense and in the full sense of the term. Representation in the minimal sense occurs when a cognitive system is capable of anticipation and can make a mistake. However, this is not yet representation in the full sense of the term, as there is no mention of the system's ability to detect errors, which for Bickhard is the condition for representation in the full sense.

9.3 The Anticipatory Function of Representation

Bickhard characterizes representation in terms of the self-maintenance function. In his conception, Cummins's function [13] is associated with the disposition towards self-preservation or self-maintenance. According to Cummins's account, a function is a certain capacity (or disposition) of a complex system whose parts are causally interconnected. The distinguished capacity of a system analyzed functionally is the ability to sustain its continuation (remaining in a state far from thermodynamic equilibrium). This capacity comprises other sub-capacities.

Thus, the capacity of any part of an organism or process occurring within it is hierarchically related to other sub-capacities; its sub-capacities contribute to the basic capacity of an autonomous organism, which is the sustaining of its continuation. For example, the function of the kidney is to filter blood, which ultimately is essential for the survival of the animal organism [14]: 74). This basic capacity is realized because one process or part functionally presupposes that another process or part effectively fulfills its function. The functional presupposition between function A and function B means that the performance of function A presupposes that function B is being fulfilled. In other words, function A cannot be performed unless the consequence of the performance of function B is maintained or arises [2].

The functions performed by various parts or processes in the organism are thus hierarchically connected by the relation of presupposition: the functions of one part presuppose the realization of the functions of other parts. And so, the function of blood filtration cannot be performed by the kidney unless the blood has the function of circulating in the circulatory system, and the function of blood circulation cannot be performed unless the heart has the function of pumping it. Functional presuppositions have certain conditions for success, which partially occur within the organism and partially in the external environment:

> The successful serving of a function for some organism by a sub-process or part of that organism may depend on other conditions holding. Such other conditions can be of various kinds and take various forms. They may, for example, be environmental and ambient, such as that the atmosphere contain sufficient oxygen. [2: 558]

The conditions for the occurrence of functions, including those external, support the performance of other functions:

> The creation or maintenance of those conditions, therefore, is itself serving a function—the function of supporting the serving of further functions by subsequent processes and parts. [2: 558]

The conditions upon which the exercise of a function depends resemble Millikanian Normal conditions—both are conditions for the occurrence of a function. In Millikan's framework, these conditions also appear hierarchically, and the criterion for their distinction is the adaptability of traits. In the interactive model, the criterion is dynamic functional organization. Therefore, the exercise of a function depends on processes occurring both within the system and in the environment. These processes determine whether the presupposed function will occur at all: when favorable conditions are present, the presupposed activity is functional; otherwise, it is dysfunctional. For instance, the blood-filtering function performed by the kidney contributes to self-maintenance, and hence the system possesses this function.

The function of representation is a special kind of self-maintenance function. It is an anticipatory function that involves indicating possible actions or preparations for generating indicators of future interactions. Bickhard refers to these indicators of internal outcomes of future system activities, which are far from thermodynamic equilibrium, as "anticipations." An example of such an anticipation could be an indicator of the sugar level in a bacterium's body. The sugar level, which in itself is not a representation, is functionally available within the organism, meaning that it influences the organism's subsequent activities. If the sugar level is sufficient, the organism does not need to search for sugar; if it is not, the organism must search for sugar until its hunger is satisfied. Bickhard [15] emphasizes that the mentioned indicators must pertain to internally available outcomes, otherwise, they would have to be first perceived and likely represented, which in turn would lead to an infinite regress.

Indicators of internal activity outcomes can be links in the chain of successive activities: a subsequent activity may depend on the success of the previous one. Thus, the further search for sugar depends on the sugar level after consuming a fruit, for example. If the organism engages in further searching, its organization presupposes a failure in the form of an insufficient sugar level. Let's assume that a bacterium is searching for sugar, but in the meantime, its hunger is satisfied. As Bickhard argues, in such a case, the anticipated state of the organism is falsified because the pursuit of further sugar sources is unnecessary. It depends on an indicator that is no longer true.

Content emerges implicitly as a result of the system's anticipatory abilities. Anticipation implicitly defines content, meaning that a functional relationship occurs within the system: if an activity were to occur at a given time under certain conditions, the system would indicate this activity as possible.

> If the frog detects an opportunity for tongue flicking and eating via a factual detection of a fly, then there is a functional relationship in the frog system that *would* have indicated that tongue flicking and eating possibility *if* a fly *had* been detected—and that *conditional relationship* for creating such an indication is functionally present in the frog even when there is no fly in that location—even when the conditional is counterfactual. That is, there must be conditional preparednesses to set up indications of interaction possibilities, even when the conditions are not satisfied and so the indications are not activated. [2: 571]

This idea can be expressed in slightly different terms: a single indicator of a certain possible action under conditions C at time t_1 is an internal state that is produced as a result of the occurrence of conditions C at time t_1 and enables the system to perform action A.

Functional presuppositions, that is, the functional conditions for the occurrence of actions in the system, determine the content of indicators. When representation occurs in the minimal sense, the content consists of indicators of the success of a possible action at a given time and place ("the action here and now").[4] Indicators are characterized by adequacy in the same sense as the actions undertaken. The test of an indicator's adequacy is the success of the presupposed action under external conditions relative to the cognitive system. Thus, the test of the adequacy of the indicator "catch here and now" is the actual catching of the object at time t1 under conditions W, which the acting cognitive system is directed towards. It is therefore clear that such understood content has conditions of fulfillment, and thus Bickhard resolves the difficult problem of content. It is worth noting that implicit anticipation is not metarepresentational, as representations are conceived in terms other than representational. Hence, this concept remains in accordance with the desideratum of the naturalization of intentionality (II.2).

Bickhard, unlike Millikan, does not emphasize the role of correspondence for representation.[5] He equates the incorrectness of representation with the failure of the action, making failed actions that are the result of correct representations impossible. However, the frog might not catch the fly if at the same time a meteorite strikes the Earth. The failure of this action then has no bearing on the adequacy of the representation. Generally speaking, even with correct representations, we have no guarantee of the success of the presupposed action.

Bickhard emphasizes the role of temporality in his model. Actions form a temporal sequence in which indicated possibilities of future actions are connected with future activities. For example, a possible action of a representing frog, such as sticking out its tongue, points to the more distant possibility of eating a fly.

Thus, the function of representation contributes to the self-maintenance of the organism. The organism has the function of implicitly indicating its possible actions when they are performed in accordance with the indication and contribute to the self-maintenance of the system. It has a dysfunction in representation when it has a dysfunction in implicitly indicating future states of the organism. This occurs when the functional presupposition of the next action is not true. The presupposition of an action can be falsified by its results, but the organism may not detect the falsehood. It is important to note that Bickhard does not equate the logical value detected by the cognitive system with the logical value of the presupposition as such. As a result, the cognitive system can imperfectly recognize the adequacy of its representations: an action can succeed even though it is based on false presuppositions. However, Bickhard does not allow for the situation where an action fails as a result of a correct representation.

Bickhard and Christensen's *prima facie* synchronic conception of function is inconsistent with Millikan's diachronic conception of function regarding the normativity of explanations using the concept of mental representations. Millikan agrees with Davidson, asserting that intentionality cannot arise without history (I wrote about the Swampman thought experiment in the previous chapter).

---

[4] Indication in the most minimal sense does not require a representation of the object, let alone the context in which the object appears. The differentiation between object and context emerges at higher levels of the representational system's development.

[5] Gładziejewski [16] makes a similar claim about Bickhard.

Bickhard and Christensen [17] and Campbell [14] place great importance on the conceptual distinction between exercising a function and having a function, considering having a function as secondary to its exercise. Proponents of interactivism use the term "having a function" to refer to the ascription of an etiological function, and "exercising a function" to refer to the current manifestation of a trait to which we attribute a function—be it etiological or systemic. In a given system, something must be able to exercise a function in order for us to say that it has that function. Bickhard distinguishes his conception of function from the etiological in this way.

Bickhard and Christensen criticize the historical approach to function, arguing that history does not affect how a function is exercised. According to the synchronic function conception, the Swampman can be attributed actual representations because it has current abilities conditioned by its structure and organization.

The Swampman thought experiment, as interpreted by Bickhard, aims to undermine the etiological concept of function, which emphasizes the importance of historicity. However, one can provide a similar example that challenges the synchronic conception, which prioritizes causal efficacy for the self-maintenance of the cognitive system. This is exemplified by representations that fall under the category of inconvenient truths, such as the truth about a patient's incurable disease, which can harm the patient in their fight against the illness (the *nocebo* effect). The etiological concept of proper function does not exclude the possibility that instances of representations may sometimes be currently detrimental to the system. According to the etiological conception, such beliefs are not problematic, as function is attributed only to the mechanisms of representation generation, and individual representations have a derived proper function. Therefore, one does not have to claim that this mechanism is dysfunctional, while the representations can still be true. Meanwhile, according to the synchronic conception, inconvenient truths are dysfunctional because they currently do not contribute to the self-maintenance of the system and thus harm it. Bickhard cannot apply the same strategy as Millikan, claiming that the function is attributed to the mechanisms that generate and receive representations. This is because currently, these mechanisms also do not contribute to the self-maintenance of the system, even if they were adaptively beneficial in the past.

It is true that it is primarily the exercise of function, rather than its possession, that speaks to the role it plays in an organism. This is the essence of the primacy of exercising a function over possessing it, as long as we are interested in the current explanation of the cognitive system's operation. In other situations, which often arise in scientific explanations, the exercise of function becomes part of a broader explanation, where adaptiveness, and thus the history of natural selection, also plays a significant role. For example, history helps explain why something is a feature of a given organism, by virtue of which it has the function of representation. For instance, a frog has the function of representing flies as food, in part because it has a visual apparatus that allows it to observe dark moving shapes. The claim that the cognitive system can exercise a function without possessing it would not be true.

Examples such as the Swampman and inconvenient truths indicate the limitations of both conceptions of function in explaining actions using the notion of mental representations. Representational explanations can pertain to various aspects of cognitive processes and activities, and are explained differently depending on the epistemic interest. For instance, in explaining the behavior of a frog catching a fly, a systems biologist need not refer to the history of frogs, whereas an evolutionary biologist is likely to do so.[6]

The most convincing, therefore, seems to be a pluralistic stance that combines the diachronic concept and the synchronic concept. In the synchronic concept of function, a necessary condition for being a representation is having the appropriate structure and organization, which indeed emerge in the evolutionary history of the organism. For this reason, it seems that both conceptions of function can coexist within a broader conception of functional explanation of representations, taking into account both current functioning as well as possessing functions. In a pluralistic approach, both conceptions of function complement each other.[7]

Regardless of whether pluralism necessitates a unification of the conceptions of function or not, the historical explanation can be supported by referring to the current causal structure. This allows for a broader explanation. For this reason, the pluralistic conception of function also seems closer to scientific practice. For example, an evolutionary biologist may refer to ethology to explain why something is an element of the communication code of bees, while an entomologist, who is interested in why bees perform the same dance at a selected time, is satisfied with referring to the current functioning of bees.

9.4 Minimal Interactive Model of Representation

In the interactive model, representing involves anticipating possible interactions of an individual cognitive system within its environment. The distinguishing features of this model among others are: interactivism, individualism, and anticipatory nature.

Cognitive systems that are ascribed the function of representing within this model actually exist in nature. The minimal requirement is that the described systems have a sufficiently complex biological organization, which would allow them to be ascribed the capability of anticipating the outcomes of their possible actions. In other words: this system can be ascribed an anticipatory organization.

Anticipatory organization is a type of functional organization characterized by appropriate complexity, gradability, and hierarchy. Anticipatory capabilities emerge in sufficiently advanced cognitive systems. Thanks to these capabilities, such systems can make choices as environmental conditions change, in order to correct or compensate for them [2]. Anticipatory organization is gradable, as the anticipatory capabilities allow for a spectrum of cognitive systems to be established—from the most

[6] More on the explanation of frog behavior in this chapter.

[7] Pluralism can be understood in various ways. For instance, Godfrey-Smith [18] argues that distinct concepts of function merely complement each other, while Kitcher [19] believes that they constitute a unified concept of function. The issue of functional pluralism is broad and goes beyond the scope of this book, but see [20].

minimal, which can maintain their own existence, to the most sophisticated, which use abstract representations, language, and mathematics. The latter usually have many possibilities for interaction, enabling them to sustain themselves even in highly diverse environments.

Anticipatory organization is hierarchical in the sense that internal interactions or the organism's actions are interconnected. This means that each individual action within the organism contributes to a functional whole. For example, the actions within a frog's organism contribute to the function of representing a fly as food, which includes focusing attention on the fly, tongue protrusion towards it, and ultimately consuming the fly, all of which constitute a functional whole. Moreover, the hierarchy of anticipatory organization allows these actions to be divided into more or less fundamental, actual capabilities of the organism to represent. The fundamental processes occurring in the organism are motor in nature, and their function is to indicate its future possible movements. Built upon these capabilities are higher sensory, cognitive, and specifically human linguistic abilities [14]. By considering different capabilities within a certain hierarchy, Bickhard's concept meets the desideratum of universality (I.2) as well as the desideratum of a multi-level nature (I.3).[8]

In the minimal interactive model of representation, temporality is limited to anticipating one's own actions in a given environment. Minimal cognitive systems do not create a network of interactive possibilities. The sequence of actions within a given pattern of action does not change. For example, a bacterium swims towards sugar or turns away when it fails to detect an increasing concentration of sugar. It is an organism so simple that it does not have access to the adequacy of its own indicators, and therefore to the content of its own representations.

9.5 A Comprehensive Interactivist Model of Representation

A comprehensive model of representation is a minimal model enriched by the ability to construct a complex network of interactive possibilities using detected representational errors. While learning through the acquisition of associations may occur even in minimally representational systems, they do not engage in learning from errors; this, however, is characteristic of more complex systems. It constitutes not only a fundamental criterion for being an actual representation but also a key desideratum for an adequate naturalistic concept of representation.

The ability to detect the adequacy of representation exists because the system can choose at least two possible actions. Detecting the adequacy of one of them involves, for example, detecting inconsistencies between its results and anticipation. For instance, a frog has indicators of possibility for projecting its tongue to the left or to the right. If the action of projecting the tongue to the left ends contrary to the frog's expectations, a failure (since there is no food to the left of the frog), then the frog

[8] Bickhard Describes individual development in a constructivist manner, much like Jean Piaget [21]. A more detailed discussion can be found in Bickhard's work [22, 2].

may attempt to project its tongue to the right. The cognitive system has the ability to detect the adequacy of functional presuppositions, that is, whether the occurrence of states of affairs that are conditions for the success of its actions are met when it is capable of detecting the outcomes of presupposed actions. It is important that not only the possibilities of actions are indicated but also the intermediate internal states (or processes) that lead to them. Only the actual state of affairs ultimately determines the success of a given action. As Bickhard asserts, the anticipatory organization of the system "does not require any representational phenomena in order for it to exist or to come into existence" [2: 570].

The system's ability to detect the adequacy of representations is thus a necessary condition for representation. Only such a system can learn from mistakes. Learning through trial and error enables organisms to create new sequences of actions. A frog serves the function of representation when it has an anticipatory function that consists of indicating possible actions in a given environment. The anticipatory organization of the frog is complex enough for the animal to have indicators with content such as 'food here now' (functional presuppositions) informing it about the possibilities of eating in the presence of a fly. The success of the performed action, such as lunging for food in the form of a fly, determines the adequacy of the content contained in the functional presuppositions. A hungry frog lunges at a fly but fails to catch it. Despite extending its tongue towards the fly, the frog does not reach it. After multiple attempts, the frog may jump and finally catch the fly, making the internal indicator of success for the leap towards the fly true (in the case of a miss, it would be false).[9] In other words: multiple attempts allow it to correct the error. As a result of experience, the frog may learn to choose situations that allow for more effective action.

The comprehensive model thus concerns more complex systems, which usually have additional biological goals beyond survival. Such systems are characterized by the ability to create new sequences of possible actions: actions are indicated (e.g., taking a drink from the fridge) that are conditioned by other actions (e.g., opening the door), which are conditioned by yet other actions, and so on. The organization of such conditional actions can create a very complex network. This network constitutes the organism's situation knowledge [2].

Bickhard's model differs from Dretske's approach when it comes to learning. The former has a certain advantage. In Dretske's concept, representing consists of indicating objects or features from the external environment, in particular, tuning or adapting the organism to the environment. In Bickhard's conception, the opposite is true. Since representing serves the purpose of anticipating possible interactions of the system with the environment, and in particular regulating its activity, the organism, by predicting its own actions in a given environment, can adapt them to the changing environment. Thus, the organism can learn to utilize its error in such a way as to improve the quality of life and become independent of the surroundings.

---

[9] Contrary to Campbell [14], a frog is capable of adaptive learning from mistakes through reinforcement.

The interactivist model is an action-oriented model *par excellence*. Interactions are constitutive for representation, as the content of representations depends on the functional presuppositions of actions. Consequently, the organism has cognitive access to the adequacy of its representations, which it can recognize in the form of outcomes of activities. This distinguishes Bickhard's approach from Millikan's. In this respect, Bickhard's model most fully incorporates the connection between representation and action among the concepts of representation discussed so far.

9.6 Bickhard's Desiderata. Is Bickhard an Internalist?

The assumptions of the interactivist model of representation apply to every cognitive representational system. Bickhard provides his own list of desiderata for an adequate concept of mental representations. These desiderata are:

1. System-detectable error;
2. Functionally accessible content;
3. Functional comparability of what is represented;
4. Content remaining in internal relations;

   and

5. Emergent representation.

Desideratum 1 reiterates Bickhard's key criterion for being a representation, which is the system's ability to detect its adequacy. The interactivist model of representation states that functional presuppositions (arising through anticipation) are falsified when they do not proceed as the system previously anticipated. Desiderata 2–4 concern the content of representations, which is functionally accessible to the system. The content of representations is comparable with other contents and possibly states of the system (desideratum 3), as representations have conditions of satisfaction. Desideratum 4 follows from desideratum 2, which requires cognitive access to representations, which according to Bickhard necessitates that content remains in internal relations to its vehicles (functional presuppositions). Desideratum 5 requires the naturalization of representations, that is, demonstrating how representations could have arisen from what is not a representation. Representational phenomena emerge from non-representational phenomena both diachronically: as new properties in history, since they concern systems with anticipatory organization, and synchronically—representational properties pertain to the system as a whole, but not to all its components.

Emergence is a type of inter-level ontological relation. If intentionality is an emergent property relative to physical properties, then on one hand, the concept of intentionality can be explicated in terms other than intentional, and on the other hand, one cannot perform a simple microreduction of the intentionality of a given representation to its components. Microreduction is understood as the reduction of systems (states, properties, processes) treated as wholes to their (or their states, properties, processes) components [23].

The justification for the claim that contents are to remain in internal relations to their vehicles is as follows. Content remains in internal relations because functional presuppositions always arise within the anticipatory organization of the cognitive system. The source of content is the cognitive system, which has access to it. However, representation becomes semantic only as a result of the organism's interaction with the environment. As such, content is wide, though it is also characterized by internalism: it always arises solely within the cognitive system. In relation to their reference, representations maintain external relations, which are not necessary for being a representation.

Thus, the source of content is the cognitive system, and this content depends on the functional organization of the system, which in turn depends on interactions with the environment. Importantly, it is not individual functional presuppositions that implicitly determine the content of representations, but rather their network. Therefore, content takes the form of implications, where the success of action 1 depends on the success of action 2. For example, a frog receives a signal reaching its retina caused by the stimulus of a small black dot moving (on the frog's horizon). The success of the frog's action depends on whether functionally presupposed actions connected by relations will occur. The content of one of many anticipations, such as lunging at a fly, could take the form of the statement: "I'm about to eat." The success of the performed action determines the adequacy of the content contained in the functional presuppositions, and this is dependent on the environment, particularly on whether there is an edible fly or, for instance, an inedible bumblebee on the frog's horizon.

It might seem that Bickhard is a proponent of internalism and adopts the internalist thesis of local supervenience of content on the microstructural properties constituting a network of internal indicators. However, these networks are closely related to the environment, as they are shaped in interaction with it. Bickhard's model is therefore not consistent with the thesis of local supervenience on the organism's microstructure, because the content in this model is wide, being connected with the environment. Certainly, however, it has an internalist aspect, since the content of representation is that which is cognitively accessible to an individual organism. Meanwhile, one might question whether the conception of an individual organism is so obvious.[10] In the case of strongly coupled biological systems, where distinguishing individual organisms is impossible, Bickhard should speak of representations available to the entire system, rather than to an individual organism.

---

[10] In evolutionary biology and the philosophy of biology, there is an ongoing discussion about the individuation of organisms. While earlier theories of evolution focused on the natural selection of organisms, there is now an increasing emphasis on allele dynamics. More radical philosophers of biology question whether the definition of an organism as a separate entity is applicable. The phenomenon of natural selection and adaptation indicates that sometimes such individuation is simply impossible. This applies not only to various kinds of symbiotic systems (e.g., termites) but also to single complex biological systems [24, 25, 26]. Even distinguishing a biological individual is problematic [27].

The apparent similarity to internalism may seem even more striking because Bickhard's conception of communicative processes does not involve the transmission of a representation between a sender and a receiver. It is questionable whether Bickhard's model adequately describes social representations, such as in the case of the waggle dance of bees. The waggle dance is an example of representations whose content arises from the coordination between individual systems. Representation does not occur at the individual level here. Moreover, it is not clear whether Bickhard's conception would manage cases of this kind, where the autonomy of the individual unit is not the overarching goal of the distributed biological system (a group of bees), due to which they have evolved representational abilities. It might seem at most that the biological goal of a bee group is to maintain its autonomy. Even assuming that the cognitive system could be (or is) a group of bees performing the dance, it is unknown whether the group's autonomy is the primary goal or merely derivative with respect to the autonomy of the individual units forming the group. Thus, Bickhard's approach seems—contrary to Millikan's more liberal approach—too narrow. Millikan's conception of representation seems to describe social representational processes much better.

In Bickhard's view, words in language are not representations themselves; they are operators acting upon the mental representations of their recipients. There is no sharing of meaning in the literal sense: whether of individual representations or entire narratives. It is worth noting that this is also the case in [28], where Millikan adopts a radically individualistic approach to concepts (which she calls "unicepts"). Therefore, one could say that the waggle dance of bees—an example of a basic mode of signaling among animals—are operators on the representations of individual insects. However, this solution has the shortcoming that all cases of signaling and communication would require the existence of representational abilities in the signal recipients. This seems odd, as this kind of signaling is evolutionarily very early. It is not obvious whether simple biological communication indeed requires not only minimal representation but also representation in the full sense of the word—complete with error detection. One might worry that any coordination of social actions would turn out to be representational, even in very simple organisms, to explain behaviors typically not considered in terms of mental representation but rather in terms of tropisms or taxes. In such an interpretation, Bickhard's approach would be too broad, as would Millikan's teleosemantics, as I have shown in the previous chapter.

9.7 Summary

In the interactivist model, the role of the relative autonomy of the individual cognitive system is emphasized. The operations of the cognitive system and its internal organization are of significance. However, the content is wide, as the adequacy of representation is a matter of objective dependencies in the world, and the content remains in internal relations with functional presuppositions shaped in interaction with the environment. The individualistic aspect of the model is particularly manifested when the cognitive system does not produce detailed representations of the environment but only its own motor states, and representation is reduced to anticipating one's own actions.

Anticipatory representations are the fundamental type of representation, and others emerge from them. The interactivist model aptly describes representational processes related to implicitly anticipating one's possible actions. It is closely connected with the dynamics of the individual biological cognitive system, which must continuously undertake actions to sustain life.

In summary: the interactivist model is effective when organisms anticipate their own actions. However, it seems doubtful whether this model can satisfactorily account for all types of representations, such as social representations.

At the same time, Bickhard's approach seems too broad. Since the structure of representation is omitted in the model, it does not consider semantically utilized correspondences [16]. Thus, it is difficult to attribute even minimal representational abilities to a bacterium, as it lacks a mechanism for distinguishing relevant features of the environment from the rest. Millikan avoids this consequence because, in her view, correspondences play a key role in determining the content of representations. Moreover, in Bickhard's model, the erroneousness of representation is reduced to the failure of action based on it, yet such action can fail even with a completely accurate representation.

Bickhard's model of representation meets almost all the desiderata from the introduction. There may be doubts as to whether it fulfills the desideratum II.3.1 regarding productivity. However, productivity is certainly not excluded, since even in the minimal model, organisms have the capacity for the implicit representation of sequential actions. Bickhard, however, does not develop this issue further.

References

1. Bickhard, M.H.: Representational content in humans and machines. J. Exp. Theor. Artif. Intell. **5**(4), 285–333 (1993). https://doi.org/10.1080/09528139308953775
2. Bickhard, M.H.: The interactivist model. Synthese **166**(3), 547–591 (2009). https://doi.org/10.1007/s11229-008-9375-x
3. Buckner, C.: A forward-looking theory of content. Ergo: Open Access J. Philos. **8**, 37 (2022) https://doi.org/10.3998/ergo.2238
4. Fodor, J.A.: A theory of content and other essays. MIT Press, Cambridge (1992)
5. Davidson, D.: What thought requires. In: Problems of rationality, pp. 135–151. Oxford University Press, Oxford (2004)
6. Davidson, D.: The problem of objectivity. In: Problems of rationality, pp. 3–19. Oxford University Press, Oxford (2004)
7. Harman, G.: Thought. Princeton University Press, Princeton, NJ (1973)
8. Maturana, H.R.: The tree of knowledge: the biological roots of human understanding. Shambhala Publications, Boston-New York (1992)
9. Maturana, H.R., Varela, F.: Autopoiesis and cognition: the realization of the living. D. Reidel Publishing Company, Dordrecht (1980)
10. Varela, F., Thompson, E., Rosch, E.H.: The embodied mind: cognitive science and human experience. MIT Press, Cambridge (1991)

11. de Jaegher, H.: Participatory sense-making: an enactive approach to social cognition. Phenomenol. Cogn. Sci. **6**(4), 485–507 (2007)
12. Paolo, E.D.: Autopoiesis, adaptivity, teleology, agency. Phenomenol. Cogn. Sci. **4**(4), 429–452 (2005)
13. Cummins, R.: Functional analysis. J. Philos. **72**(20), 741–765 (1975). https://doi.org/10.1007/978-1-4614-7004-5
14. Campbell, R.J.: The concept of truth. Palgrave Macmillan, New York (2011)
15. Bickhard, M.H.: Genuine representation in artificial systems. In Antoniou, G., Slaney, J. (eds.) Advanced topics in artificial intelligence. Lecture notes in computer science, pp. 27–38. Springer, Berlin, Heidelberg (1998) https://doi.org/10.1007/BFb0095038
16. Gładziejewski, P.: Action guidance is not enough, representations need correspondence too: a plea for a two-factor theory of representation. New Ideas Psychol. **40**, 13–25 (2016). https://doi.org/10.1016/j.newideapsych.2015.01.005
17. Bickhard, M.H., Christensen, W.: The process dynamics of normative function. Monist **85**(1), 3–28 (2002)
18. Godfrey-Smith, P.: Philosophy of biology. Princeton University Press, Princeton (2014)
19. Kitcher, P.: Function and design. Midwest Stud. Philos. **18**(1), 379–397 (1993)
20. Bielecka, K., Marcinów, M.: Mental misrepresentation in non-human psychopathology. Biosemiotics **10**(2), 195–210 (2017). https://doi.org/10.1007/s12304-017-9299-2
21. Piaget, J.: The construction of reality in the child. Transl. M. Cook. Basic Books, New York (1954)
22. Bickhard, M.H.: Cognition, convention, and communication. New York (1980)
23 Poczobut, R.: Między redukcją a emergencją: spór o miejsce umysłu w świecie fizycznym. Wydawnictwo Naukowe, Wrocław (2009)
24. Bouchard, F., Rosenberg, A.: Fitness, probability and the principles of natural selection. Br. J. Philos. Sci. **55**(4), 693–712 (2004). https://doi.org/10.1093/bjps/55.4.693
25. Bouchard, F.: Causal processes, fitness, and the differential persistence of lineages. Philos. Sci. **75**, 560–570 (2008)
26. Turner, J.S.: Extended phenotypes and extended organisms. Biol. Philos. **19**, 327–352 (2004)
27. Clarke, E.: The problem of biological individuality. Biol. Theory **5**(4), 312–325 (2010). https://doi.org/10.1162/BIOT_a_00068
28. Millikan, R.G.: Beyond concepts: unicepts, language and natural information. Oxford University Press, New York (2017)
29. Anderson, M.L., Rosenberg, G.: Content and action: the guidance theory of representation. J. Mind Behav. **29**(1–2), 55–86 (2008)
30. Anderson, M.L., Rosenberg, G.: A brief introduction to the guidance theory of representation, In: Proceedings of the 26th Annual Conference of the Cognitive Science Society, ed. K. Forbus, D. Gentner, T. Regier, Erlbaum (2004)

Chapter 10
Coherence-Based Account of Representational Error

In this chapter, I propose my own approach to system error. The structure of the chapter is as follows: I analyze misrepresentation in terms of etiological dysfunction (Sect. 10.1). In my approach, I utilize Millikan's producer-representation-consumer system and Bickhard's conception of representation in action, though I do not consider anticipation a necessary condition for representation. I argue that only the capacity to detect representational error enables us to attribute cognitive accessibility of its own representations to an organism, as it has the ability to detect its own error. I also outline the conditions a system must meet to detect an error (Sect. 10.2). First, I present my coherence-based approach to system error detection through selected examples (Sect. 10.3), followed by the utilization of detected errors in learning processes (Sect. 10.4). I compare the proposed conception with previously discussed accounts of representational error (Sect. 10.5) and demonstrate applications of this conception to various types of representations, using examples from mind and brain sciences (Sect. 10.6). Finally, I examine potential objections to this conception and to teleosemantics in general (Sect. 10.7).

10.1 Dysfunction of Representing, or Representational Error

Elsewhere, I maintain that the concept of function should be approached pluralistically, meaning a hybrid concept of function that combines historical and system functions should be adopted [1]. However, since the paradigmatic examples discussed do not require such an expansive concept of function, I define system error in terms of dysfunction within Millikan's framework (see Chap. 8). The historical concept of function is sufficiently rich to encompass not only the nature of system error but also its role in learning. To address the emergence of entirely new mental structures (as

K. Bielecka, *What is Misrepresentation?*, Studies in Applied Philosophy, Epistemology and Rational Ethics 71, https://doi.org/10.1007/978-3-031-84375-4_10

in the Swampman thought experiment), a more complex, hybrid concept of function that includes system function would be necessary.

I conceive of the representational function—following Millikan—as an etiological proper function. A representation token has its referent (e.g., a fly) and content (i.e., conditions of correctness), as well as a physical vehicle (a state of the nervous system realizing a state of representation). Correctness is determined by the actual state of affairs, but the consumer has the ability to recognize the adequacy of a representation (though infallibility is not assumed).

A representational error is a type of dysfunction in the representational system. Dysfunction occurs when a proper function is not fulfilled because some of the numerous Normal conditions for that function are absent. An organism (or part thereof) has a representational dysfunction when its capacity for representation is preserved. Misrepresentation involves a discrepancy between the content of a representation and the represented object. For instance, a misrepresentation of X occurs when the represented object X is not present.

10.2 System-Representational Error

A system-representational error is a dysfunction that an organism recognizes as an error. The phrase "as an error" might suggest that an organism can only recognize its error if it has a representation of the error, that is, it knows something is a misrepresentation. I will demonstrate that error recognition occurs through a response to inconsistencies between representations without the necessity of meta-representation. Moreover, this is the fundamental method of detecting representational errors, which does not require advanced cognitive abilities. First, I will describe the conditions a system must meet to detect errors; then, I will analyze how a system can utilize its error. The first set of conditions must be met by any error-detecting system, while the use of detected errors in learning is just one of many ways cognitive systems employ detected errors in cognitive processes. The example of representation in a frog serves as the model.

The capacity to utilize representational errors logically follows the ability to recognize them. Recognizing representational errors involves detecting inconsistencies among pieces of information, at least one of which is semantic (meaning it has conditions of satisfaction), or distinguishing erroneous representations from correct ones. The necessity of inconsistency for the detection of a representational error is precisely why I term this conception "coherence-based." In philosophical tradition, this term is used for the coherence theory of truth, according to which the truth of a judgment or a set of judgments is reduced to their coherence. However, the coherence-based conception of representational error does not rely on the premises of the coherence theory of truth; it is logically independent from it.

The fact that cognitive systems can detect an error by identifying inconsistencies in representations does not imply that undetected representations are true or accurate. It is entirely possible that a cognitive system may never detect the erroneous nature of its actually inaccurate representations. In the latter case, the organism's ability to regard certain information as more reliable compared to other information is required, thereby enabling the differentiation of more reliable (correct) representations from less reliable (erroneous) ones. In the cognitive processes of an organism, some information is prioritized and processed as more reliable (in essence: accepted as veridical), which facilitates the distinction between more reliable (correct) and less reliable (erroneous) information. Only an organism capable of discerning less reliable information from more reliable information can differentiate erroneous representations from correct ones.

The functional ability to differentiate less reliable information from more reliable information is the capacity to consistently regard certain environmental information as more reliable, as opposed to other information. This may be evident in a particular kind of perceptual constancy, such as the size constancy of an object in motion (approaching or receding from us). Size constancy means that an object does not alter its perceived size as it moves closer or farther away, even though its retinal image enlarges or shrinks. Certain relationships of a given object with other perceived objects, such as the obscured surface of an object directly behind it, remain constant. Hence, the variability of an object's retinal image size during motion is considered by our visual apparatus as more reliable information than the invariability of an afterimage in the eye, which remains constant. Consequently, an afterimage is not perceived as an object. Moreover, an afterimage can be disregarded when observing objects simultaneously within our field of vision: information about the presence of an afterimage is deemed less reliable than the presence of objects moving relative to us.

The ability to regard certain information as more reliable logically precedes the ability to represent and also constitutes an organism's capacity to differentiate between erroneous and correct representations, which in turn conditions the ability to recognize representational errors. The differentiation between more reliable and less reliable information can only occur after detecting an inconsistency between those pieces of information.

Therefore, the sufficient condition for a visual representational mechanism to register (or represent) an inconsistency of information is the differentiation of an object and its separation from other objects (distinguishing figure from ground) and the spatial localization of the object (egocentric or allocentric). Conversely, the sufficient condition for a representational mechanism to eliminate such inconsistency is the discernment of information reliability, for instance, through perceptual constancies. The elimination of this inconsistency presupposes the aforementioned ability to differentiate between less reliable and more reliable information.

Every system (organism) that possesses the following structure can recognize an error within the system. The system must 1) be equipped with at least one representation and either another representation or a piece of information deemed inconsistent with the previous one (e.g., a pain stimulus that need not be a representation in the full sense of the term). Error correction requires that one of the pieces of information or representations be more reliable than the other, with both being simultaneously accessible. A system capable of recognizing a representational error must also have 2) a representational mechanism capable of registering inconsistencies between these representations, which may be evidenced by a response to the inconsistency. The foundational assumptions of the coherence-based model for recognizing and correcting representational errors are presented in Table 10.1.

Representational errors may occur quite frequently (analogous to other types of dysfunctions discussed by Millikan). The representational mechanism of a frog capable of recognizing an error is primarily a mechanism for representing in a normal manner. A frog has a mechanism for normally representing, for example, flies, when its representational capabilities (of flies) are adequate given the prevailing conditions (the presence of flies).

The ability to normally recognize, for instance, bumblebees or flies, is possible when the representations have content accessible to the organism, such as a frog. Thus, the frog's representational mechanism can recognize an edible fly when the producer subsystem, in response to moving black dots, generates a representational state that, in turn, triggers a consumer reaction, initiating a behavioral sequence leading to the consumption of the fly. This state is representational because the consumer system can further verify the accuracy of this state, for example, by comparing it with another concurrent representational state (e.g., an auditory state indicating the presence of a potentially painful stinging insect). The existence of the consumer's proper function to respond to certain types of representational states generated by the producer also results in a mapping between these types and reality.

Mappings have an adaptive significance and thus serve the frog's functions. There can be numerous different mappings between the states of the frog's nervous system and reality, but only those that can be verified, i.e., those in which the cognitive system can recognize an error, have a representational character. Hence, I endorse Bickhard's emphasized notion of a system-detectable error.

However, contrary to Bickhard, I do not assume that all other types of representations emerge from anticipatory representations. The frog's access to representations of bumblebees and flies is ensured by the presence of a producer-representation-consumer system that utilizes and can compare both representations. The producer subsystem generates these representations in response to stimuli, such as flying flies or other similarly appearing objects, while the consumer subsystem uses them to direct the frog's motor activities. Moreover, since such a system plays an adaptive role and is present in every frog or toad, it can be said to have a proper function

Table 10.1 Coherence-based Model for Recognizing and Correcting Representational Errors

The ability to recognize representational error A presupposes: • a producer and consumer system capable of registering inconsistencies in representations, • the availability of representation A and representation/information B, which is considered inconsistent with A.
To correct a misrepresented A, one needs: • the ability to recognize error A, and its manifestation requires: • the ability to discern the reliability of information A and B.

that ensures reproductive success and the continued existence of this system in the population, thereby preserving the capacity for representation within the population.

10.3 Examples of Detecting Representational Error

Consider the detection of representational errors using the example of a frog that has at least two representations: a fly and a bumblebee, as well as a representational mechanism capable of responding to inconsistencies between these representations. This description is a simplification and may not directly correspond to biological reality. For simplicity, I also omit an important issue from my perspective, which is the existence of a certain system (repertoire) of interconnected representations. I adopt this requirement from Millikan, but since it does not play a significant role in the example under discussion, I will omit it. For the purposes of this example, let's assume that amphibians can compare information from different senses. Inconsistency arises, for example, when a frog has two incompatible representations, visual and tactile, of the same object (see: Fig. 10.1). The frog receives a visual stimulus (moving and elongated) that evokes a representation of a fly. Consequently, the motor system (the consumer of this representation) triggers the tongue to shoot out toward the object.

However, tactile stimuli from the tongue evoke a representation of a bumblebee instead of another fly representation. This leads to the detection of inconsistency, and thus a representational error. If the frog did not react more strongly to tactile than to visual stimuli, it could not immediately discard the fly representation but only detect that one of the two representations is incorrect. For the sake of analysis, let's assume that tactile information is indeed more reliable than visual, making the fly representation erroneous.

The ability to respond to inconsistencies in the content of representations and information is a necessary condition for recognizing representational errors. Some cognitive systems are also capable of representing these inconsistencies, but this is not a necessary condition for misrepresentation.

Fig. 10.1 Illustration of the example of detecting representational errors

10.4 Utilizing Errors in Learning

A condition for misrepresentation is the organism's functional ability to distinguish certain information as more or less reliable. This kind of ability is a necessary condition for all learning, particularly for reinforcement learning, which is fundamental to adaptation and, according to contemporary neurobiological hypotheses, is closely associated with verifying the accuracy of an organism's predictions [2]. Representational errors can indeed be exploited in reinforcement learning. Through learning, an organism acquires the ability to compare the outcomes of its actions with the rewards it anticipates.

Suppose the frog has an innate mechanism that allows it to differentiate between representations of a fly and a bumblebee and can learn through reinforcement to distinguish between these representations. The producer subsystem reacts to a stimulus on the frog's retina, initially generating a fly representation, followed by the consumer initiating a sequence leading to the insect's consumption. However, consuming an insect does not result in a reward if the frog captures a bumblebee, and thus—through reinforcement learning—after repeated attempts to catch a bumblebee, it begins to differentiate the state generated in the presence of a bumblebee from that generated in the presence of a fly. This differentiation could occur, for example, according to the neurobiological learning mechanism proposed in the actor-critic architecture literature, where the action-taking system (controlled by the consumer) is overseen by a system that checks its results (the reward system) [3].

The actor-critic architecture posited in computational neuroscience is compatible with the proposed account of representation. Assuming it underlies reinforcement learning, any organism capable of discerning information and learning through reinforcement can exploit its learning errors. This occurs when a producer repeatedly generates a representation that elicits an incorrect consumer response, resulting in punishment or absence of reward for the organism. The organism then begins to differentiate the state induced by this representation from the state produced in the presence of a representation that triggers the correct, rewarding response. From an evolutionary perspective, the organism learns to distinguish between what is beneficial, such as representing an edible item, and what is less beneficial, like representing an inedible item. Thus, a coherence-based conception of the system-detectable error allows us to assert that representations underpin reinforcement learning in the actor-critic architecture (Fig. 10.2).

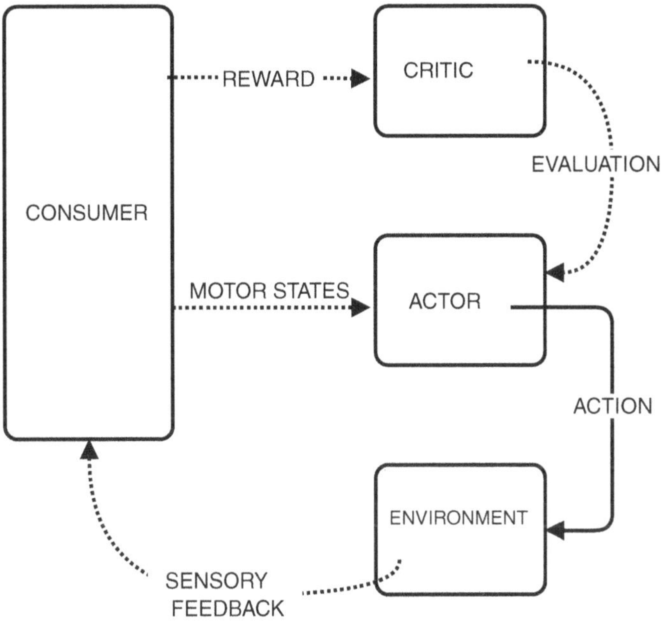

Fig. 10.2 The actor-critic architecture

10.5 A Proposed Conception of System Error and Other Teleosemantic Conceptions of Representational Error

The conception of representational error presented here is a preliminary and minimal proposal for an account of mental misrepresentation in the spirit of teleosemantics (for an application to error correction in singing birds, as studied by [4], see [5]). It is timely to summarize the fundamental similarities and differences between this account of representational error and the teleosemantic conceptions discussed thus far.

I most notably distance myself from Dretske's approach to representational error. Although he acknowledges the possibility of misrepresentation as a crucial element of a satisfactory account of representation, thereby meeting desideratum II.3.2 on the potential for misrepresentation, his content theory is insufficient to endow a representational system with the capacity to assess representation accuracy. Nevertheless, Dretske has influenced other teleosemantics scholars to consider the necessity of misrepresentation in evaluating the adequacy of a naturalistic representation concept.

In the proposed conception, I incorporate elements of Millikan's representation theory, which aligns with the evolutionary theory of natural selection. I address representational error as a dysfunction in the context of Millikan's historical proper function. However, I contend that the notion of proper function need not be exclusively historical; biological function also has a system dimension. In other words, a coherence-based conception can rest on a hybrid biological function theory (integrating diachronic and synchronic perspectives).

I concur with Bickhard's assertion that a conception of mental representations should include the cognitive system's own error recognition, at least in some instances. Yet, my approach to detecting representational errors markedly diverges from Bickhard's. Firstly, I frame system-detectable errors within Millikan's producer–consumer framework. Furthermore, I reject Bickhard's premise that recognizing representational adequacy requires the system's capacity to anticipate its representations as either adequate or inadequate. Among other benefits, this approach allows the coherence-based conception to accommodate various representation types, particularly those that do not necessitate anticipation. Bickhard also equates detecting inadequacies with action failure detection [6], however, as I have demonstrated, failed actions can also result from accurate representations. In a coherence-based conception of the system-detectable error, the criterion for error detection is the identification of representational or informational inconsistencies.

10.6 Application of Conceptions to Examples in Mind and Brain Sciences

Prior to this chapter, I deliberately withheld examples from scientific practice that I had previously used to assess the adequacy of various mental representation accounts, as different teleosemantic approaches converge on similar analyses of these examples. While other theories, both internalist and externalist, diverge from the dominant scientific understanding of mental representation, teleosemantics exhibits a unique alignment with this concept. This alignment is not achieved by simply describing mental representations in scientific practice, but by providing a comprehensive conception of mental representations that not only meets the meta-desideratum of consistency with scientific research but also offers normative guidance on how to use and evaluate mental representations in a scientifically informed way. Unlike some other teleosemantic accounts, such as those proposed by Dretske, Millikan, or Bickhard, my approach does not trivialize the concept of mental representation, thus satisfying Ramsey's requirement. Through examples of rombix, mental imagery, and cognitive maps, I will demonstrate how the conception under discussion diverges from other teleosemantic approaches. The differences between teleosemantic conceptions stem from the adoption of a particular conception of function and boil down to the treatment of representational content; the representational vehicle remains the same—it is a state of the nervous system.

Rombix

As demonstrated, Lewicki's research aims to engender a concept, hence a representation, of rombix in the subject. The test for assessing knowledge of the rombix representation hinges on the ability to differentiate rombices from non-rombices. The experimenter presents various cards to the subject, who categorizes them as rombices or non-rombices.

In Dretske's view, concepts are quintessential mental representations. The function of representation is one of indication, which Dretske later [7], following Millikan, interprets in terms of an etiological proper function. Thus, the function of representing rombices is a proper function derived from the representational function of the entire conceptual apparatus. This apparatus is marked by its capacity for learning, and therefore, according to Dretske's definition of conceptuality, it is conceptual. The content of the individual ROMBIX concept is a complex indicator determined by regular covariation between the concept's vehicle in the subject and the rombix figure, based on laws or probable dependencies resulting from the learning process. This can be schematized as follows: a neurological state (d) possesses a given property (g) if and only if a card (w) has the property of depicting a specific figure (f).

Millikan's approach diverges from Dretske's in her analysis of mental representations within a producer–consumer system. This system is endowed with a proper function of representation. Conceptual representations, whose vehicle is language, find their place within this system. However, the producer and consumer are distinct within the system, allowing the content of representation and the actions it elicits to remain separate. The producer is thus the perceptual system, and the consumer is a suitably complex inferential-motor system.

Within her etiological framework of representational function, Millikan considers not only the selection of indicators—single or complex, as the later Dretske does—but also the selection of mapping rules themselves. Hence, the content of an individual representation (which is a derivative function relative to the proper function of representing rombices)—in this case, the concept (or unicept) of ROMBIX—emerges from a selected mapping rule under Normal conditions; in the example, these pertain to the correct functioning of the perceptual and inferential-motor systems. The content of the ROMBIX concept is thus a mapping between the concept and the figure depicted in the image. Millikan thus permits representations of objects that are typically nonexistent, which could pose a problem for Dretske. However, in the case of rombix, this is of little consequence, as it is an artificial concept.

Bickhard posits that the system function of representation is anticipatory. Conceptual representations, and more broadly abstract ones, are incorporated into his model of representation, where content is tested through action, including reasoning action. The content of the ROMBIX concept consists of functional presuppositions related to action, such as distinguishing rombices from non-rombices. The cognitive system accesses the content of its own representations by accessing the representational vehicle, which is intrinsically linked to the content. In this manner, the content is necessarily connected to the vehicle.

The coherence-based conception of the system-detectable error is grounded in the etiological proper function of representation and Millikan's producer–consumer framework. I also draw on Bickhard's notion of system-detectable error. Thus, a necessary condition for a representational mechanism in the subject to have the capacity to register (or represent) informational inconsistencies is the perceptual-conceptual ability to distinguish between a rombix and a non-rombix. The necessary condition for a representational mechanism in the subject to eliminate such inconsistency is the ability to discern levels of information reliability, for example, by inferring from the experimenter's reactions.

The mechanism for representing a figure is capable of recognizing, for instance, a rombix, when the producer subsystem generates a representational state in response to a drawing presented by the experimenter. This state triggers a consumer response, initiating an inference sequence that leads to a behavioral sequence resulting in an affirmative or negative response to the tester's question about the presence of a rombix in the drawing. This state is representational because the consumer system can further verify its accuracy, for example, by comparing it with another concurrent representational state (such as a memorized drawing of the figure that appeared earlier or was refreshed through inferential processes). The existence of the consumer's proper function to respond to certain types of representational states generated by the producer also creates a mapping between them and reality. In such a complex representational system as that of humans, it is challenging to speak directly of the adaptive role of mappings. However, the exploitation of error can be adaptively beneficial in a broader sense, as it is motivated by a cognitive reward that ultimately manifests in the brain's dopamine release.

In more complex learning scenarios concerning higher-order representations, such as learning the concept of ROMBIX, we encounter both an innate mechanism in the system capable of automatically recognizing and eliminating representational error and learning from mistakes.

Mental Imagery

In Ernst Mach's experiment, mental images of the right figure are compared with the actual perception of the left figure. I will show that teleosemantic theories address the representation of currently non-existent objects in various ways.

In Dretske's theory, mental images are not examples of mental representations, as mental representations require concepts; they are certainly instances of cognitive representations. It is noteworthy that in his theory of representation, there is no place for addressing their structural properties, although the selection of indicators logically entails the selection of mappings. However, Dretske does not discuss the selection of the mapping rules themselves, complicating the representation of currently non-existent objects, including mental images. Dretske would have to consider an image as a quasi-percept to deal with the discussed example. Then, the representational function, or the selected indication function, would occur between a neuronal state (d) with property (g) whenever the visual cortex state (w) has visible properties (f).

In Millikan's theory—as in Dretske's later work—the representational function for mental images has been selected. Unlike Dretske's theory, the appropriate mapping rules have been selected, whether through natural selection or learning. An organism with such a function can operate on non-existent figures. Thus, Millikan's approach allows for the manipulation of figures in the imagination.

Bickhard's anticipatory model also does not encounter problems with imagery. In his theory, there is room for operations on abstractions, such as rotating geometric figures. For Bickhard, the representational function involves anticipating a state of affairs, such as rotating a given figure to the right. The content is tested in action, for example, by rotating the figure in a specified direction.

In the coherence-based theory, mappings are selected similarly to Millikan's approach, while representational adequacy is—following Bickhard—recognizable by the system in action. Thus, rotating a figure "in the mind" may or may not correspond to a drawing on the subject's right side.

Cognitive Maps

Cognitive maps serve as another example of structural representation [8]. The hypothesis that rats can represent inaccurately appears credible, thus allowing us to more confidently propose that cognitive maps are representational in nature (although one might posit that something is a representational mechanism without yet having discovered the error-correction mechanism, such a hypothesis would be weaker than if the mechanism were identified). Empirical evidence suggests that a rodent's sensory stimuli are utilized to correct errors in current or future locations [9], and the presence of an EEG negativity signal indicates the rodent's awareness of an error.

Consider an experiment where a rat, familiar with a particular maze, is transferred by researchers to a very similar but foodless maze. The rat, now in a slightly altered maze, retains the cognitive map of the previous one.

Dretske's later work allows us to posit a selected representational mechanism tasked with indicating the rat's spatial location. In Millikan's framework, mapping rules are also selected.

The rat can recognize its error when it discovers the absence of food at the maze's end, possessing a structural representation in the form of a cognitive map ([10]). This aligns with the conception of a system-detectable error as proposed by both Bickhard and the coherence-based approach.

Rats possess a representational mechanism capable of reacting to inconsistencies within these representations. An inconsistency arises when a rodent, recalling the location of food in the old maze, is surprised by its absence in the expected spot. This triggers sensory stimuli that inform the rat's brain of the food's absence (I). The rat's representational mechanism then registers the inconsistency between the previous cognitive map and the current location information, prompting the release of dopamine to the rat's brain, with the recognition of inconsistency manifested by a behavioral surprise reaction (II). Upon approaching the represented object, it becomes apparent (III) that the maze is devoid of food.

A rat recognizes its error only when it can respond to inconsistencies, namely when it learns that there is no food in that maze section and consequently avoids it in the future.

10.7 Objections to the Proposed Account of Misrepresentation

The conception of representational error sketched in this chapter is preliminary and vulnerable to numerous objections. I will limit the discussion to three primary criticisms: the accusation of vulgar pragmatism, the charge of adopting verificationist premises, and the alleged mischaracterization of normativity.

The vulgar pragmatism objection

The charge of vulgar pragmatism is often leveled against teleosemantics, as seen in critiques by [11] and Burge:

> Biological functions are functions that have ultimately to do with contributing to fitness for evolutionary success. Fitness is very clearly a practical value. It is a state that is ultimately grounded in benefit of its effects for survival for reproduction (...) Explanations that appeal to biological function are explanations of the practical (fitness) value of a trait or system. But accuracy is not *in itself* a practical value. Explanations that appeal to accuracy and inaccuracy such as those in perceptual psychology are not explanations of practical value, or of contributions to some practical end.
>
> Deflationist theories are part of a long, failed tradition of assimilating truth and accuracy to contribution to practical success, and falsity and inaccuracy to practical failure. Error need not be a failure or frustration of any independently identifiable biological function. Representational success need not fulfill any biological function. [12: 301]

Teleosemantics is criticized for equating veridicality or truth with functional properties: the correctness of a representation is identified with meeting a biological need or benefit, while the incorrectness of a representation is associated with failing to meet such a need or benefit. This leads to undesirable consequences similar to those of the vulgar pragmatic conception of truth. According to this view, truthfulness or adequacy is reduced to utility:

> You can say of it then either that 'it is useful because it is true' or that 'it is true because it is useful.' Both these phrases mean exactly the same thing, namely that here is an idea that gets fulfilled and can be verified. True is the name for whatever idea starts the verification-process, useful is the name for its completed function in experience. True ideas would never have been singled out as such, would never have acquired a class-name, least of all a name suggesting value, unless they had been useful from the outset in this way. [13: 204]

Indeed, such objections challenge those teleosemantic theories that attribute functions to individual representations rather than to the mechanism of their creation [14, 15]. The conception of misrepresentation I propose avoids this mistake because the biological function is attributed to the entire producer-representation-consumer system. In this sense, the benefit pertains to the functioning of the whole producer and consumer system, not to individual representations or even specific representational functions.

Consider the example of the frog. Reducing incorrectness to a lack of benefit would occur if the adequacy of an individual representation of a fly were determined by the frog's satiation. Yet a hungry frog may accurately represent the fly it is pursuing. Moreover, in the model I propose, a function is ascribed to the entire motor-sensory system of the frog, and the incorrectness of a representation does not deprive the producer–consumer system of its function. Thus, the concept of the system-detectable error does not exclude situations where, despite unmet hunger, the frog's sensorimotor system operates flawlessly. It is biologically advantageous because the biological function is partly determined by a suitable history, not solely by current processes.

Consequently, a coherence-based conception of misrepresentation allows for the existence of beneficial misrepresentations, provided they are instances of individual representations. For example, catching bumblebees is not beneficial, but it is advantageous for the frog's sensorimotor system to operate in such a way that it is adequate in relation to reality. Vulgar pragmatism, however, would not permit such a scenario, as it deems all errors to be detrimental.

While I reject the vulgar pragmatism claim that truth reduces to the benefits of action, I do not deny the significant role of action in cognition and in the pursuit of truth. This does not imply that a system has an infallible criterion of truth. The epistemic value of pragmatism lies in the cognitive system's access to mappings it can recognize and produce through broadly defined action, including thinking. Only those mappings potentially accessible through the system's action inform the accuracy of its representations. The mappings available to the system are objective in that their correctness or error depends on the system's capabilities and its environment, as per Millikan's concept of locally recurrent natural signs (later called information signs, or infosigns; see [16]). In general, a coherence-based conception of system-detectable error presupposes the assumptions of the correspondence theory of truth, since the correctness of representations is determined by selected mappings between the representation and its environment. The pragmatic element is evident in that organisms can learn mappings through broadly defined action, the outcomes of which usually serve as a fallible indicator for the system of whether its representation is correct or incorrect. Even actions based on correct representations can fail, thereby not benefiting the individual organism. Thus, the connection with action is not as direct as vulgar pragmatism suggests. Notably, the coherence-based conception also includes a coherentist element, as the consistency of information is a condition for recognizing the adequacy of representation.

The critique of an improper approach to the normativity of functions

The issue of the normativity of meaning is extensively debated in philosophy [17, 18]. Representations are normative in that they can be judged as correct or incorrect, allowing for evaluative discourse. Naturalistic, and by extension teleosemantic, conceptions of representation are criticized for their flawed approach to normativity [12, 19, 20, 21–23]. The critique posits that teleosemantic theories erroneously reduce the normativity of representation to biological or evolutionary normativity, thereby misrepresenting the traditional understanding of representational normativity, this is typically linked to higher forms of representation and involves adherence to certain social standards to which we are beholden. Daniel Hutto and Erik Myin assert, "There is a fundamental mismatch between the kinds of norms associated with truth or accuracy and those associated only with biological functionality" [19, 20: 112]. However, their book lacks an argument to support this claim. A comparable assertion is made by Burge, who speaks of "a *root* mismatch between representational error and failure of biological function" [12: 301]). Stich contends, "natural selection does not care about truth, it cares only about reproductive success" [23: 62].

This objection, however, is misguided because it rests on an overly simplistic understanding of natural selection and the role of functions in teleosemantics. Since functions are ascribed to traits on which natural selection is predicated, discussions of biological function pertain not so much to natural selection per se or adaptation in general, but rather to specific organismal traits with adaptive roles. Natural selection operates precisely because of these traits. If the objection were valid, one could also argue that evolution does not favor accurate vision mechanisms but is based on reproductive success. Under this view, however, vision could not be biologically functional, as biological functions—according to the objection—would serve only reproductive success (and that for the entire population, not just individual organisms). Representation, on the other hand, involves a mapping rule between an organism's properties and what is represented, and the function of representation was selected precisely for its representational trait. Just as with vision, where one can discuss certain standards of sharpness or depth that affect the adaptive value of sight, so too can one discuss the accuracy or inaccuracy of representations. It is not, as Stich might have it, that representation directly serves reproduction, though it is undoubtedly a trait that typically contributes indirectly to reproduction. This occurs precisely through mappings that enable an organism to adequately understand its environment.

The critique of an improper approach to the normativity of functions is a double-edged sword. There may be doubts about the meaning of the concept of normativity, which supposedly applies only to linguistic and social representations. The claim that this type of normativity is fundamentally different from biological normativity requires an explanation of why this is the case. However, it appears that the normativity associated with human communication could be equally applicable to bee communication. To refute this, one would need to demonstrate the fundamental difference between the normativity of human and animal communication.

Demonstrating this difference will be particularly challenging for those philosophers who seek to naturalize intentionality. In characterizing the norms governing linguistic and social representations, they cannot rely on non-naturalized concepts of content; thus, they cannot discuss the content of, for example, conventional norms consciously adopted by language users. They would need to develop a concept of normativity that is naturalistic and pertains to phenomena such as animal communication or epigenetic reproduction mechanisms [24].

The Verificationism Objection

The proposed theory might be criticized for adopting verificationist assumptions, meaning that the criterion for the meaningfulness of a representation is empirical verification: according to verificationism, meaningful statements are all and only those that can be empirically verified [25]. An example of this approach is the purely causal theory of content, which has been criticized by [11], among others. On this account, individual representations are verified one by one.

The proposed model indeed requires a significantly weaker form of verification that does not necessarily encompass every representation. A cognitive system need not have the capacity to verify all representations generated by the producer; it must be able to verify at least some of them. Verification need not be empirical, as the cognitive system can detect inconsistencies among pieces of information that do not originate from sensory systems. This does not imply that representations lack truth conditions, but rather that they may remain technically unverifiable by the organism. For animals, this may not be as consequential as it is for humans, where it is not possible to ascertain whether verification occurs, though we do not wish to deny them truth conditions (e.g., the mental imagery of individuals with schizophrenia). Thus, the resemblance of the coherence-based model to traditional verificationism is quite superficial; it does not entail any form of logical empiricism.

10.8 Conclusion

The coherence-based conception of the system-detectable error is a preliminary conception of mental representations that meets all the desiderata I have advocated for. Some of the solutions adopted in this conception stem from naturalistic premises, while others draw from the premises of teleosemantic theories, enriched by additional assumptions of specific solutions, particularly those associated with Millikan.

The meta-desideratum I concerning the relationship between the theory of representation and scientific inquiry is characteristic of naturalistic conceptions, and the proposed conception is no exception. The coherence-based conception satisfies the explanatory desideratum (I.1) and the universality desideratum (I.2) through naturalization and the biological underpinnings of teleosemantics. This conception meets Millikan's intervention condition, providing an outline of the cognitive architecture's structure, applicable across all organizational levels of the cognitive system, namely the functional producer–consumer system. In line with Millikan, I embrace the biological theory of communication and frame linguistic behaviors within its model. Since the goal of any teleosemantic conception, including the one presented here, as well as other functional naturalistic conceptions, is to furnish a conceptual framework that accurately describes the cognitive architectures of the systems under study, it can be employed to characterize both personal and subpersonal phenomena (thereby satisfying desideratum I.3).

Metadesideratum II concerning the relationship between the theory of representation and common practice is acknowledged by all naturalistic theories of representation, which postulate an explication of the pre-existing concept of representation in such a way as to account for the rationality of the cognitive system (desideratum II.1). The proposed conception of the system-detectable error is no exception.

Thus, teleosemantics ensures a link between representation and reference (desideratum II.2.1 on the determination of reference and content) and elucidates why a representation can carry content (desideratum II.2.2), while also accommodating the possibility of misrepresentation (desideratum II.3.2). It is also not an outlier among teleosemantic theories, satisfying the crucial desideratum of naturalizing intentionality (desideratum II.2). By adopting Millikan's concept of Normal conditions, I demonstrate that adequacy is a matter of the appropriate relationship between the cognitive system and its reference (described in terms of local natural information). A coherence-based conception, not being an exception among teleosemanticists, describes content and reference in such a way that they are non-accidentally related. By adopting Millikan's producer–consumer framework, I show the conditions a cognitive system must meet to determine reference. In particular, such a system must have access to its own content. Desideratum II.3.1 on the productivity of at least some representations is met because Millikan's model accommodates representational productivity. The second key desideratum—about system-detectable error (II.3.3)—is met in the discussed conception thanks to the development of a coherence-based model for system error detection. Thus, the coherence-based conception fully satisfies the desideratum for a proper concept of content (II.3).

The proposed account stands up well against common criticisms of other teleosemantic theories of representation, such as the vulgar pragmatism objection, the mischaracterization of normativity objection, or the verificationism assumption objection. Certainly, the proposed sketch of the conception requires refinement when describing many phenomena of misrepresentation, particularly higher-order representations such as hallucinations or delusions. The question of the accuracy of delusions in individuals with mental disorders is particularly challenging and widely debated [26–28]. However, this does not imply that the coherence-based conception of representational error lacks the conceptual tools needed for their description, such as Millikan's producer–consumer system, which allows for the description of representations at both the individual and social levels. For instance, by relying on my teleosemantic model of episodic memory, I account for confabulations in amnesic patients in Korsakoff Syndrome [29].

In summary, the proposed conception offers a sketch of a satisfactory naturalistic theory of mental representation. It delineates a minimal model of system error that accounts for representational phenomena without requiring metarepresentational capacities. This model is sufficiently liberal yet does not trivialize the concept of representation and meets Ramsey's requirement, suggesting that, with appropriate modifications, it could serve as a draft for a conception of representational error in specific, empirically researched instances of mental representation. However, providing suitable examples is not the sole responsibility of the philosopher but also of the scientist.

References

1. Bielecka, K., Marcinów, M.: Mental misrepresentation in non-human psychopathology. Biosemiotics **10**(2), 195–210 (2017). https://doi.org/10.1007/s12304-017-9299-2
2. Colombo, M.: Constitutive relevance and the personal/subpersonal distinction. Philos. Psychol. **26**(4), 547–570 (2013). https://doi.org/10.1080/09515089.2012.667623
3. Khamassi, M.: Actor-critic models of reinforcement learning in the basal ganglia: from natural to artificial rats. Adapt. Behav. **13**(2), 131–148 (2005). https://doi.org/10.1177/105971230501300205
4. Gadagkar, V., Puzerey, P.A., Chen, R., Baird-Daniel, E., Farhang, A.R., Goldberg, J.H.: Dopamine neurons encode performance error in singing birds. Science **354**(6317), 1278–1282 (2016). https://doi.org/10.1126/science.aah6837
5. Bielecka, K., Miłkowski, M.: Error detection and representational mechanisms. In: Smortchkova, J., Dołęga, K., Schlicht, T. (eds.) What are mental representations?, pp. 287–313. Oxford University Press (2020)
6. Gładziejewski, P.: Wyjaśnianie za pomocą reprezentacji mentalnych: perspektywa mechanistyczna. Warszawa–Toruń: Wydawnictwo Naukowe (2015)
7. Dretske, F.I.: Naturalizing the mind. The MIT Press (1997)
8. Gładziejewski, P., Miłkowski, M.: Structural representations: causally relevant and different from detectors. Biol. Philos. **32**(3), 337–355 (2017). https://doi.org/10.1007/s10539-017-9562-6
9. Ferbinteanu, J., Shapiro, M.L.: Prospective and retrospective memory coding in the hippocampus. Neuron **40**(6), 1227–1239 (2003). https://doi.org/10.1016/S0896-6273(03)00752-9
10. Miłkowski, M.: Satisfaction conditions in anticipatory mechanisms. Biol. Philos. **30**(5), 709–728 (2015). https://doi.org/10.1007/s10539-015-9481-3
11. Fodor, J.A.: A theory of content and other essays. MIT Press, Cambridge (1992)
12. Burge, T.: Origins of objectivity. Oxford (2010)
13. James, W.: Pragmatism. A new name for some old ways of thinking. Longmans, Green and Co., New York (1907)
14. Dretske, F.I.: Knowledge and the flow of information. Cambridge University Press (1981)
15. Macdonald, G., Papineau, D.: Teleosemantics: new philosophical essays. Oxford University Press, Oxford-New York (2006)
16. Millikan, R.G.: Beyond concepts: unicepts, language and natural information. Oxford University Press, New York (2017)
17. Brandom, R.: Modality, normativity, and intentionality. Philos. Phenomenol. Res. **63**(3), 1–34 (2001)
18. Posłajko, K.: Spór o normatywność znaczenia w analitycznej filozofii języka. Kwartalnik Filozoficzny **37**, 123–156 (2009)
19. Hutto, D.D., Myin, E.: Radicalizing enactivism: basic minds without content. MIT Press, Cambridge (2013)
20. Hutto, D.D., Myin, E.: Exorcising action oriented representations: ridding cognitive science of its Nazgul. Adapt. Behav. **21**(3), 142–150 (2013). https://doi.org/10.1177/1059712313482684
21. Heras-Escribano, M., de Pinedo, M.: Are affordances normative? Phenomenol. Cogn. Sci. **14**(4), 565–589 (2015). https://doi.org/10.1007/s11097-015-9440-0
22. Kripke, S.: Wittgenstein on rules and private language: an elementary exposition. Harvard University Press, Cambridge, Mass. (1982)
23. Stich, S.P.: The fragmentation of reason: preface to a pragmatic theory of cognitive evaluation. MIT Press, Cambridge (1990)
24. Korbak, T.: Scaffolded minds and the evolution of content in signaling pathways. Stud. Log., Gramm. Rhetor. **41**(1), 89–103 (2015)
25. Zabłudowski, A.: Polemiki filozoficzne: prace zebrane. Wydawnictwo Naukowe, Warszawa (2009)

26. Cooper, R.: Psychiatry and philosophy of science. Acumen Publishing, Stockfield (2007)
27. Ghaemi, S.N.: The concepts of psychiatry: a pluralistic approach to the mind and mental illness. Johns Hopkins University Press, Baltimore (2007)
28. Thornton, T.: Essential philosophy of psychiatry. Oxford University Press, Oxford-New York (2007)
29. Bielecka, K.: Confabulations in Korsakoff's Syndrome: defending an error-based account. Erkenntnis (2024). https://doi.org/10.1007/s10670-024-00857-w

Conclusion

The primary task of the book was to justify and organize the requirements that a satisfactory philosophical conception of mental representation must meet, and to develop a naturalistic theory of mental representation that provides an appropriate account of representational error. In this work, I defended the thesis that the coherence-based account of error for the system, which I have developed, fulfills this role. The justifications were provided by selecting a particular list of desiderata presented in the introduction and by analyzing the arguments in favor of competing naturalistic conceptions of mental representation. I referred to the examples from mind and brain sciences introduced in the introduction, as they are recognized by philosophers and scientists—even opponents of the notion of mental content [1, 2]—as indubitable instances of mental representation. I illustrated the discussed concepts with these examples, treating them as a kind of test of their adequacy. The analysis of concepts using scientific examples provided additional justification for the desiderata I defended. The journey through various naturalistic conceptions of representation is presented in the form of a decision tree in Appendix B.

Meta-desiderata and desiderata constitute a list of requirements for the sought-after conception of representational error. This theory was to be congruent with both scientific practice (meta-desideratum I) and common understanding (meta-desideratum II). I argued that the cognitive system's error detection in its representations is a condition for it to be capable of representing at all. The system's ability to detect its representational error is necessary for it to have access to the correctness of its own representations, which is closely related to their content.

The argumentation in favor of the list of desiderata I have adopted was largely based on a review of competing positions, which—as I have demonstrated—do not allow for an adequate description of both everyday and scientific practice. It is worthwhile to briefly outline the course of this argumentation in the individual chapters of the book. My starting point was Dennett's intentional stance. I showed that intentionality, the fundamental feature of representations, characterizes all thinking. Following Dennett, I posited that common practice provides the framework for the scientific

K. Bielecka, *What is Misrepresentation?*, Studies in Applied Philosophy, Epistemology and Rational Ethics 71, https://doi.org/10.1007/978-3-031-84375-4

concept of intentionality, and thanks to this feature, representations serve to describe, explain, and predict behaviors, in accordance with the explanatory desideratum (I.1). Contrary to Davidson, I assumed that intentionality could be attributed not only to humans but also to animals, which corresponds to the desideratum of universality (I.2). I argued for a representational account of behavior at the level of whole organisms as well as subpersonal cognitive processes, which is consistent with the desideratum of multi-level analysis (I.3). The result of Chap. 2 was a characterization of intentionality, as well as an indication of the connection between representations and rationality (desideratum II.1).

In subsequent chapters of the book, I sought a conception that would naturalize intentionality (desideratum II.2) and provide a proper account of content (desideratum II.3).

In Part I, I sought an adequate conception of representational content among functionalist positions. I analyzed simple functionalist views: single-factor—primarily Fodor's—and dual-factor—Block's—as well as Wisconsin-style externalist semantics and the naturalistic semantics of truth conditions (NTCS). In Chaps. 3 and 4, I examined internalist conceptions of representation. I demonstrated that internalism has certain advantages, as it emphasizes the internal dimension of a cognitive system's representation and highlights the role of intentionality. However, the concept of narrow content, even in Block's dual-factor conception, was unsuitable for the explication of concepts used in mind and brain sciences. These sciences describe actual cognitive processes, the understanding of which is only possible by considering features such as reference or veridicality. I concluded that internalist conceptions of representation, in any version, failed to meet the desideratum of determining reference and content (II.2.1) or the desideratum of a content vehicle (II.2.2). Consequently, they do not fulfill the overarching desideratum of naturalizing intentionality. By not explicating the concept of reference and content, they failed to properly capture the relation between intension and extension, and thus completely satisfy the desideratum of an appropriate concept of content (II.3), although they met the desideratum of productivity (I.3). In the dual-factor version, more desiderata were met, thanks to wide content.

In Chaps. 5 and 6, I discussed purely externalist and non-teleosemantic positions, which explicated the concept of reference and satisfaction conditions (the latter only pertains to the conceptions in Chap. 6), partially fulfilling the desideratum of intentionality (II.2). However, the omission of intentionality led to conceptual difficulties, such as the problem of trivialization and the disjunction problem.

In the conclusion of Part I, I stated that conceptions of mental representation should indicate determinants of both reference and content (desideratum II.2.1) and show why mental representations can bear content. The failure to meet these desiderata means that representations do not cognitively connect with reality, and it is unclear how to assess their adequacy. Therefore, intentionality cannot be naturalized within the discussed non-teleosemantic externalist conceptions (desideratum II.2).

In Part II, I analyzed teleosemantic theories, specifically: Fred Dretske's (Chap. 7), Ruth Garrett Millikan's (Chap. 8), and Mark Bickhard's (Chap. 9). Teleosemantics not only provided an account of wide content but also captured the relation between intension and extension in a manner closest to the evolutionary paradigm in biology.

Teleosemantics addresses problems that are particularly troublesome for the functionalist conceptions discussed earlier, namely Daniel Hutto and Erik Myin's hard problem of content and the disjunction problem (horizontal and vertical).

Hutto and Myin [1, 2] argue that the conceptions of Dretske and Millikan do not specify how a representation can bear content, as it is unclear how the contents of representations can be assigned satisfaction conditions (the hard problem of content). This would mean that these concepts do not meet desideratum II.2 on the naturalization of intentionality.

In Dretske's approach, representation is defined in terms of an indicator (part of the definition of an indicator reads as follows: "d is G" functionally means that "w is F"), and since indicators constitute a kind of predication (taking the form of protosentences), they—contrary to Hutto and Myin—have conditions of satisfaction. Millikan's conception, on the other hand, explained why a representation can be a bearer of content because representations connect with the world through mappings. The criticism of Hutto and Myin against teleosemantics, that it supposedly does not answer the question of why a representation can have contents, did not hit any of the conceptions discussed in this part (Dretske, Millikan, Bickhard, nor a coherence-based conception of error for the system). Thus, these concepts meet the desideratum about the bearer of content (II.2.2), and thereby the desideratum on the naturalization of intentionality (II.2).

At the same time, it should be mentioned that—as Millikan emphasizes—teleosemantics "piggybacks" on other concepts of representation:

> To the shell that is "teleosemantics" one must add a description of what *actual* representing is like. When the bare teleosemantic theory has been spent, the central task for a theory of intentional *representation* has not yet begun. Teleosemantic theories are piggyback theories. They must ride on more basic theories of representation, perhaps causal theories, or picture theories, or informational theories, or some combination of these. [3]: 66]

In other words: teleosemantics required supplementation with an additional account of content, as it does not itself determine what content is or its nature. However, it is worth noting that all the representatives of teleosemantics discussed here have supplemented it in one way or another—Dretske with the conception of functional meaning, that is, the function of indicating; Millikan with the conception of mapping rules and local natural information, and Bickhard with the conception of contents of anticipation.

Desiderata concerning the content of representations posed the greatest challenge to all naturalistic conceptions. Teleosemantics was no exception. However, Millikan's and Bickhard's conceptions seemed to come closest to fulfilling these desiderata. They described content and reference in such a way that they became interconnected in a non-accidental manner. Unlike Block's two-factor conception, teleosemantic conceptions were not a patchwork of two separate accounts of reference and content. This is because they pointed to extra-logical dependencies between reference and content. Thus, teleosemantics allowed for discussions about the adequacy of representations with respect to what the representations are about, acknowledging that only certain features of the environment are represented by organisms. Nevertheless, Desideratum II.3.1 on productivity was not met by Dretske's and Bickhard's conception. It is important to emphasize that this desideratum was not enforced by teleosemantics, but was associated with a system-focused account of representation, which is most fully present in Millikan's conception.

Even the simplest among them—Dretske's conception—solves the hard problem of content by Hutto and Myin, while Millikan also addresses the vertical disjunction problem. However, teleosemantics could not meet Ramsey's requirement of non-trivializing the concept of representation, as it conceives representations too broadly. Bickhard, on the other hand, ties the logical value of representations too closely to the success of action.

Significantly, teleosemantics enabled the attribution of content to representations that are erroneous, thus fulfilling the desideratum on the possibility of error (II.3.2). This is because, in teleosemantics, concepts such as "representing," "misrepresenting," and "non-representing" are explicated in terms of function, dysfunction, and lack of function, respectively. As a result, representational error—unlike in other conceptions discussed in this book—was not reduced to non-representation. Teleosemantic conceptions fulfilled all the desiderata I advocate for in various ways, though only Bickhard's proposal met the key desideratum on error detectable by the system (II.3.3). Teleosemantics also struggled with this. Among the accounts discussed, only Bickhard's anticipatory conception meets it. In Chap. 10, I proposed my own solution that satisfies this desideratum, without requiring the capacity for anticipation as a condition for representation. Both Bickhard's conception and my coherence-based conception of error for the system are the only ones among those discussed in this book that fully met Desideratum II.3 on the proper accounts of content. My coherence-based theory alone fulfilled all the desiderata I advocate for.

I have proposed a certain advancement of the concept of a system-detectable error in the last chapter of my work. There, I outlined my minimalist approach to system-detectable error, which is based on Millikan's producer–consumer framework, yet unlike Bickhard's model, it does not require the organism's anticipatory capabilities. I made the ability to detect informational inconsistencies a necessary condition for error detection. I demonstrated that system recognition of an error can lead to its cognitive exploitation, for instance, in reinforcement learning, which is a fundamental evolutionary form of learning.

In summary, a coherence-based account of mental representations refers to the weak notion of representational error detection—a weakness in the sense that it does not necessitate metarepresentation. This allows organisms lacking metarepresentational abilities to detect errors. The coherence-based account differs from Bickhard's model in that it does not directly link representational adequacy to the success of action, thus not ruling out incidental action success despite adequate representations. Some of the more challenging critiques of teleosemantic concepts, such as the critique of vulgar pragmatism, inadequate treatment of functional normativity, or verificationism, do not apply to the coherence-based conception.

A substantial role in the book was also played by examples of mental representations in scientific practice. I defended the thesis that a coherence-based conception of representational error is sufficient to accurately describe this practice. This is because, on the one hand, it benefits from the achievements of teleosemantics, and on the other, representational error within its scope is framed in terms of inconsistencies among representations. In Chap. 10, I detailed how, in the case of the rombix, cognitive maps, and mental imagery, the coherence-based conception proves to be effective.

The outline of a coherence-based model for errors detectable by the system requires refinement. Representational errors belong to a large group of cognitive errors that are commonly found in organisms and artificial systems [4]. Research on errors has various theoretical and practical applications [5]. For instance, engineers study errors with more practical aims in mind, focusing on their elimination, when constructing a robot, one should aim to eliminate design flaws in the project. Cognitive psychologists, on the other hand, study cognitive errors, treating them as interesting mental phenomena. There are many types of errors that occur in planning, information storage, or the execution of cognitive tasks, for which various mechanisms are responsible [5, 6]. Research on animals and individuals with mental disorders yields results that are interesting both theoretically and practically. They have a significant impact on the prevention of these diseases and are also of interest on ethical grounds [7].

A coherence-based error model emphasizes the role of error detection and excludes treating all individuals with mental disorders as irrational, even if they misrepresent to a significant degree. I hope that this model can also be applied to research on more complex cognitive processes, such as those in individuals with mental disorders. The outline of a coherence-based teleosemantic conception provides not only the conceptual foundations for minimal phenomena of representing and misrepresenting but also for more complex representational phenomena. The development of an empirical theory of these disorders, based on the assumptions of the conception, is a task not solely of philosophy, but also of science.

Appendix A
Conceptions of Mental Representation

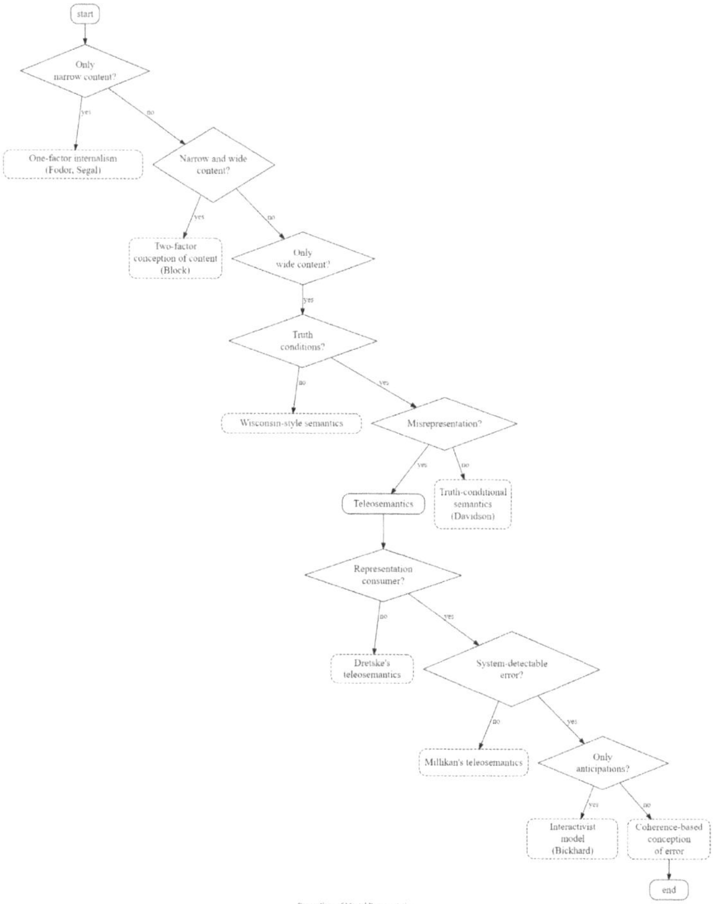

Conceptions of Mental Representation

K. Bielecka, *What is Misrepresentation?*, Studies in Applied Philosophy, Epistemology and Rational Ethics 71, https://doi.org/10.1007/978-3-031-84375-4

Appendix B
Desiderata

Meta-Desideratum I: A satisfactory theory of representation is connected with empirical research.

I.1. A satisfactory theory of representation takes into account the role of representational concepts in describing, explaining, predicting, and controlling cognitive processes and behaviors.

I.2. A satisfactory theory of representation concerns both linguistic and non-linguistic (e.g., perceptual) phenomena in humans and animals.

I.3. A satisfactory theory of representation concerns phenomena considered at both the personal and subpersonal levels.

Meta-Desideratum II: A satisfactory theory of representation can explicate the representational nature of intentional concepts in everyday and scientific practice.

II.1. A satisfactory theory of representation is consistent with assumptions regarding the rationality of cognitive systems.

II.2. A satisfactory theory of representation naturalistically explicates the concept of intentionality, that is

II.2.1. Indicates determinants of reference and content;

II.2.2. Shows why a representation can bear content.

II.3. A satisfactory theory of representation provides an adequate concept of content (including the non-contingent relationship between intension and extension).

II.3.1. A satisfactory theory of content of representation takes into account the productivity of component parts of representation.

II.3.2. A satisfactory theory of the content of representation accounts for the possibility of misrepresentation.

II.3.3. A satisfactory theory of the content of representation accounts for the capacity of the cognitive system to recognize the accuracy of the representation.

K. Bielecka, *What is Misrepresentation?*, Studies in Applied Philosophy, Epistemology and Rational Ethics 71, https://doi.org/10.1007/978-3-031-84375-4

References

1. Hutto, D.D., Myin, E.: Radicalizing Enactivism: Basic Minds Without Content. MIT Press, Cambridge (2013)
2. Hutto, D.D., Myin, E.: Exorcising action oriented representations: ridding cognitive science of its Nazgul. Adapt. Behav. **21**(3), 142–150 (2013). https://doi.org/10.1177/1059712313482684
3. Millikan, R.G.: Varieties of meaning: the 2002 Jean Nicod lectures. MIT Press, Cambridge (2004)
4. Marcus, G.F.: Kluge: The Haphazard Evolution of the Human Mind. Houghton Mifflin, Boston (2008)
5. Reason, J.: Human Error. Cambridge University Press, Cambridge (1990)
6. Kahneman, D.: Thinking, Fast and Slow. Farrar, Straus and Giroux, New York (2011)
7. Broome, M.R., Bortolotti, L., Mameli, M.: Moral responsibility and mental illness: a case study. Camb. Q. Healthc. Ethics **19**(2), 179–187 (2010). https://doi.org/10.1017/S0963180109990442

K. Bielecka, *What is Misrepresentation?*, Studies in Applied Philosophy, Epistemology and Rational Ethics 71, https://doi.org/10.1007/978-3-031-84375-4

The manufacturer's authorised representative in the EU is Springer Nature Customer Service Centre GmbH, Europaplatz 3, 69115 Heidelberg, Germany. If you have any concerns regarding our products, please contact ProductSafety@springernature.com

Printed and bound by CPI Group (UK) Ltd, Croydon, CR0 4YY
01/10/2026
02225950-0003